Green Superpowers

Green Superpowers

China, the European Union, and the United States in the Global Energy Transition

Andrea Prontera

OXFORD
UNIVERSITY PRESS

OXFORD
UNIVERSITY PRESS

Great Clarendon Street, Oxford, OX2 6DP,
United Kingdom

Oxford University Press is a department of the University of Oxford.
It furthers the University's objective of excellence in research, scholarship,
and education by publishing worldwide. Oxford is a registered trade mark of
Oxford University Press in the UK and in certain other countries

Published in the United States of America by Oxford University Press
198 Madison Avenue, New York, NY 10016, United States of America

British Library Cataloguing in Publication Data
Data available

Library of Congress Control Number: 2024942156

ISBN 9780198875284

DOI: 10.1093/9780191987304.001.0001

Printed and bound by
CPI Group (UK) Ltd, Croydon, CR0 4YY

Contents

List of Figures

List of Tables

1
Introduction

Greening international relations

The shift from wood to coal in the nineteenth century is generally viewed as the first global energy transition, a process that involves modification of the main sources of energy that satisfy the world's energy demand. Similar processes are long and complex businesses in which an energy source is usually added to the existing ones rather than immediately replacing them. Technological breakthroughs are necessary but not sufficient conditions to kick-start such transformations. Political, social, and economic factors combine—often at different times in different places—to accelerate or slow down the pace of energy transitions that have to face strong resistance: they create several losers along with potential winners. What transitions have in common is their disruptive effects, which reverberate from everyday life to the very functioning of the international system and global economy. At the dawn of the industrial revolution, coal fuelled an incredible (and intertwined) set of social changes and technological innovations that dramatically transformed societies and politics all over the world. The leading role of the United Kingdom (UK) in nineteenth-century global politics and the first golden age of globalization, which ended with the First World War, are both inextricably linked to the transition to coal. However, it was especially with the diffusion of oil in the twentieth century that the links between energy and international relations escalated. This second global energy transition from one fossil fuel, coal, to another, oil—and later also natural gas—also had an enormous impact for everyday life and world politics. Entire regions endowed with abundant cheap oil resources, such as the Middle East, became the focus of competition between the great powers. The very ascendancy of the United States as a new hegemonic leader cannot be separated from its prominence in the global oil market.

It is hardly surprising then that in the wake of the current global 'green' energy transition—the shift from fossil fuels to renewables and low-carbon technologies—scholars and policymakers alike are increasing their efforts to make sense of this fundamental change and its consequences. Scholarly

Green Superpowers. Andrea Prontera, Oxford University Press. © Andrea Prontera (2024).
DOI: 10.1093/9780191987304.003.0001

attention, while relatively recent, is moving even faster than the actual transition itself. There is no doubt that the world is still dominated by fossil fuels, whereas, with the exception of hydropower, the global diffusion of the other modern renewable sources—that is, solar, wind, geothermal energy, bioenergy, and marine energy—is still in its infancy. However, unlike previous transitions, the current one must be accelerated as rapidly as possible to decarbonize the global energy system and avoid a climate crisis that would transform life on earth as we know it. In other words, in view of the energy–climate nexus—energy-related emissions account for three-quarters of global carbon dioxide (CO_2) emissions, which make up the vast majority of greenhouse gas emissions (GHG) and are the primary drivers of climate change—the urgency of the green transition has no equal in the past. To achieve the 2015 Paris Agreement goals and limit global warming to well below 2°C—preferably to 1.5°C—compared to pre-industrial levels, fossil fuel usage must be drastically reduced in the next few decades. By 2050, according to the International Energy Agency Net Zero Emissions Scenario, global coal, oil, and natural gas consumption must decline respectively by 90%, 75%, and 55% compared to 2020 levels (IEA 2021a). At the same time, the deployment of renewables—particularly solar and wind energy, which are the most promising sources for the decarbonization of the global energy system and mitigating climate change (IPCC 2022)—will require an impressive scaling up. By 2050, almost 90% of global electricity generation should come from renewables, with solar photovoltaics (PV) and wind growing from the current 10% to nearly 70% (IEA 2021a).

Since the early 2010s, under the headings of 'geopolitics of energy transition', 'geopolitics of renewables', or 'geopolitics of low-carbon technologies', a burgeoning literature has developed to study emerging dynamics and possible longer-term scenarios for the low-carbon world (e.g., Scholten and Bosman 2016; O'Sullivan, Overland, and Sandalow 2017; Scholten 2018; IRENA 2019; Overland 2019; Hafner and Tagliapietra 2020; Scholten et al. 2020; Van de Graaf and Sovacool 2020; Bazilian et al. 2020; Bordoff and O'Sullivan 2022; IRENA 2022a; Scholten 2023; Tocci 2023; for a review see Vakulchuk et al. 2020; Blondeel et al. 2021; Colgan and Hinthorn 2023). To be sure, the bulk of the research on energy and international affairs still revolves around oil and natural gas (Buzan and Lawson 2015; O'Sullivan, Overland, and Sandalow 2017; Vakulchuk et al. 2020; Colgan 2021). After all, we live in a fossil fuel world and the precise path and timing for getting rid of these sources is far from certain. The 2022 invasion of Ukraine by the Putin regime, empowered by oil and gas revenues, and Russia's use of the energy weapon to punish Europeans and win friends around the world was just another (rough) remainder of this simple reality; as were the tensions spreading into international energy markets in the wake of the October 2023 Hamas terrorist attacks and the subsequent Israeli military response, which increased concerns for Middle East stability. However, under this more traditional layer of international politics, an entire new set of trends has gradually but steadily materialized.

The literature has focused specifically on three overarching (and interconnected) themes: the structural differences between the geopolitics of fossil fuels and renewables; the possible impact, in terms of cooperation and conflict, of the diffusion of renewables on international relations; and the potential winners and losers in the green energy transition. The most striking distinction is based on the fact that, unlike fossil fuel resources, renewables are more evenly distributed around the globe and are (of course) inexhaustible resources. This implies that traditional confrontational issues—such as resource scarcity, security of external supplies, energy dependency, and weaponization and cartelization of energy resources—hardly apply to a low-carbon world. This should relieve many sources of tension that have characterized much of the twentieth-century global politics and continue to affect international affairs today. The very distinction between consumer and producer states makes less sense in a global economy run by renewables. Geography will still matter, as some territories have better renewables potential in terms of solar radiation, wind speed, freshwater, land availability, and so forth. New energy partnerships and green trade links will be developed. Some countries might become exporters of renewable energy in the form of biofuels, electricity, or green hydrogen, that is, hydrogen obtained from water using electrolysis powered by renewables. Other countries might balance domestic production with a quota of imports if this appears feasible and convenient. International energy trade, however, will be less significant in a low-carbon world. According to the International Energy Agency, it will drop from $1.5 trillion today to $0.9 trillion in its 2050 Net Zero Scenario, despite the significant growth in the global population and economy expected by that date (IEA 2021b). Moreover, energy trade in a low-carbon world will be more regionalized, possibly reinforcing tendencies towards fragmentation of international economic exchanges. Electricity—which will play a greater role in a decarbonized energy system—is much harder and more expensive to transport over long distances than fossil fuels. The same applies to hydrogen, another energy carrier that is very important for the decarbonization of hard-to-abate sectors (i.e., heavy-duty transportation such as trucking, shipping, and aviation and heavy industry such as cement, steel, and chemicals) and for green industrial development. Like natural gas, hydrogen would be traded primarily on a regional scale. For countries with geographical proximity, hydrogen could be transported by pipelines, whereas over longer distances it could be liquefied or converted into ammonia and then transported by ship. Both these energy carriers—and the trade in 'green' electrons or molecules that they allow—have their own geopolitical implications (e.g., Van de Graaf et al. 2020; Westphal, Pastukhova, and Pepe 2022). However, for technical reasons, electricity can hardly be used in the manner of an energy weapon. Moreover, countries will always have more opportunities to replace imports with domestic production; a choice that was obviously constrained in the fossil fuel world.

As renewables are more abundant, the technologies to capture, store, and transport them, such as solar panels, wind turbines, batteries—which are essential for energy storage, for allowing large-scale deployment of renewables in the electricity sector, and for moving new generations of electric vehicles (EVs)—electrolysers, and electricity grids will become very important in the low-carbon world. A green technology race has already started, with major powers realizing how leading in these sectors and related supply chains will be crucial for economic well-being and national security in the twenty-first century. New geopolitical maps of potential winners and losers—beyond the usual suspects such as energy importers and oil-exporting countries, respectively—have been drafted according to different combinations of resource endowment, technological, economic, and institutional capacities that might translate into more (or less) influence in the emerging low-carbon world (e.g., IRENA 2019, 2022a). Finally, those minerals (e.g., cobalt, copper, lithium, and nickel) and rare earth elements (e.g., dysprosium, neodymium, and terbium) critical for manufacturing low-carbon technologies have come under the spotlight. Permanent magnets for wind turbines or EVs require rare earth elements (REEs) such as neodymium and dysprosium. Batteries, for EVs or for stationary systems, typically use lithium, but also nickel and cobalt. Solar energy technology, especially PV, requires REEs and nickel, which are also important for electrolysers for hydrogen. In addition, all these technologies and electricity grids use large amounts of copper. The transition to renewables therefore means a shift from a fossil fuels-intensive to a minerals-intensive energy system, since low-carbon technologies use a larger amount of such inputs than their traditional counterpart (IEA 2021b). For example, EVs require six times the minerals inputs of conventional cars, while an onshore wind power plant requires nine times more mineral resources than a gas-fired power plant of the same capacity. Most REEs are in fact not so rare (USGS 2022; IEA 2021c, 2023a). They are geologically abundant, although they are found in dilute concentration and are often difficult to separate, making mining and processing challenging, capital intensive, and problematic from an environmental viewpoint. Technological innovations, on the other hand, can reduce the centrality of some critical minerals, which, unlike fossil fuels, can also be recycled. Moreover, unlike fossil fuel disruptions, which (often) affect all consumers, causing energy security risks with shortages and price spikes, disruptions in the supply of critical minerals affect the construction of new green assets, but not the existing energy prices or flows. However, new dependencies in the area of low-carbon technologies manufacturing and in relation to those minerals and REEs essential for their realization can create frictions among states. The concentration of production or processing in one or a few countries could make supply chains vulnerable to market power, logistical risks, and geopolitically induced disruptions, especially through trade restrictions (e.g., IEA 2022a, 2023a; Leruth et al. 2022). This is why, in parallel with the literature focused on the above-mentioned overarching themes, a second strand of research has emerged, which studies (on a more micro-level) trade disputes over specific low-carbon technologies and/or critical minerals and REEs, or the international political economy and the politics of certain low-carbon technologies and related supply chains (e.g., Lewis 2014; Hughes and Meckling 2017; Kalantzakos 2017a, 2020; Sattich et al. 2021). In this vein, however, other works have focused on the new collaborative industrial partnerships and governance structures of the global green business, pointing to complementarities and cooperation agreements that have favoured innovations and manufacturing advances around the world, as well as to the interactions between internal and external green industrial and energy policy (e.g., Hughes and Quitzow 2018; Nahm 2021).

Overall, the current global energy transition has triggered a welcoming 'greening' of international relations. With this term I intend two interlinked trends. The first refers to the way in which the literature is keeping pace with the changing reality on the ground. Second, what is very interesting and promising about this emerging literature is its strong interdisciplinarity and the consequential blurring of the internal/external divide. This is well reflected in the efforts to combine insights from traditional international relations perspectives with a diversity of disciplines, ranging from energy policy and innovation studies to transition studies and green industrial policy. These two trends are producing a type of scholarly innovation that is similar—and this is no accident—to that prompted by the 1970s oil crisis, which reinvigorated the international political economy (IPE) tradition and international relations theories at large (e.g., Van de Graaf et al. 2016; Kuzemko, Lawrence, and Watson 2019). Energy issues, again, are driving an exciting period for those interested in world politics and global political economy. However, the downside of this race to keep pace with the evolving reality on the ground is the low level of conceptual development that still characterizes this new area of study. Scholars are often more interested in designing scenarios, discussing opportunities and challenges, and suggesting policy recommendations than developing analytical frameworks and theories. This is understandable in the light of what is at stake for countries around the world and the planet itself. As the current energy transition is widely affected by policy decisions, taking part in this conversation is certainly important (for a brief review of the positions of political scientists and other scholars on the matter, see Colgan and Hinthorn 2023). In this book, however, I adopt a different perspective. My main goal is to develop the conceptual landscape of this emerging field. Put simply, I want to reinforce those analytical building blocks that will help to further advance the 'greening' of international relations.

In particular, I want to close the existing gaps in the literature regarding the 'greening' of foreign energy policy by addressing a number of simple questions that so far remained unsolved (and unasked). What is a 'green' foreign energy policy? What are the main drivers of the major powers' green foreign energy policies? What are the policy tools at the disposal of governments to achieve their objectives in this field? What kind of power resources can they mobilize to achieve these objectives? How is it possible to map and explain the differences and similarities between the major powers' green foreign energy policies? Which green power strategy is more effective in the face of the current global energy transition? The burgeoning literature on the geopolitics of renewables, low-carbon technologies, and energy transition hardly engages with these questions. As mentioned above, the literature oscillates between offering wider pictures on the matter and more detailed empirical studies on specific aspects of the international political economy and politics of renewables. What is missing, however, is the middle level represented by the study of green foreign energy policy. This area of research has been overlooked and undertheorized in the existing literature. This shortcoming extends to crucial actors in the global energy transition, namely China, the United States (US), and the European Union (EU). These three players produce more than half of all global CO_2 emissions (China, approximately 30%; the US, 14%; and the EU, 9%; although the per capita emission rankings still place the US in the first position) and are the world's first, second, and third largest energy and fossil fuel consumers, respectively (Ritchie, Roser and Rosado 2023). At the same time, however, they are the front runners in the green technology race. China has unexpectedly emerged as a leader in the deployment of renewables and in many low-carbon technology supply chains. As a country with a very high dependency on imported oil and gas, and a rising energy demand, Beijing (in general) benefits in several ways from the shift away from fossil fuels. However, from a climate change perspective and a strategic viewpoint, China's leadership is increasingly contested by other major powers. The EU, as a large oil and gas importer with minor and declining fossil fuel reserves, is also a potential winner in the energy transition. It has invested a great deal in building its reputation as a global green leader, with only mixed results, however. Unlike China and the EU, the US is an important oil and gas producer and, thanks to the shale revolution, exporter. Fossil fuels have traditionally been an essential ingredient of Washington's international posture and influence. Despite the fact that the US was initially a leader in renewable energy research and development, in the decades after the end of the Cold War it became more of a follower in the global energy transition. Only recently, and with significant internal contradictions, have renewables and low-carbon technologies (re)gained attention in Washington's strategic thinking and global action. Although an increasing number of books touch upon diverse aspects of these three major powers' domestic and international energy (and climate) policy (e.g., Bernell and Simon 2016; Wurzel, Connelly, and Liefferink 2016; Geri and McNabb 2017; Kalantzakos 2017a, 2017b; Finamore 2018; Gang 2019; Lin 2019; Andrews-Speed and Zhang 2019; Opsal 2019; Minas and Ntousas 2019; Kuteleva 2021; Siddi 2023; Tocci 2023), a comparative and comprehensive study of their green foreign energy policy and green power strategy based on a coherent and structured analytical framework is still lacking. In this book I develop such original framework and I apply it to analyse, compare, and explain the recent efforts of China, the EU, and the US in this area. I therefore illustrate the links between the domestic green approaches that these actors are promoting and their external actions regarding renewables and low-carbon technologies supply chains. Accordingly, I argue that the different political economies of Chinese, US, and EU energy policies encapsulate distinct patterns of state–market relations, which influence the structural context of their respective green foreign energy policies and green power strategies. I also draw attention to the limits and potential of Chinese, European, and US green power strategies. As the shift from fossil fuels to renewables unfolds in the midst of a rising great power competition, these issues are fundamental to an understanding of global politics in the twenty-first century.

In the rest of the chapter, I present and discuss data to sketch the contours and drivers of the global green energy transition and to highlight the role of

China, the EU, and the US in that transition. I go on to explain why these three actors can be regarded as green superpowers which deserve particular attention in the current context. Finally, I explain the structure of the book in more detail.

Renewables and low-carbon technologies: climate change, geopolitics, security, and economic imperatives

As mentioned, we still live in a fossil fuel world. Energy transitions work by incremental additions rather than abrupt substitution of existing energy sources. The rise of renewables in the past few decades has occurred in parallel with the expansion of fossil fuels, which still account for more than 80% of world primary energy consumption, which, in turn, has increased threefold since 1970 (Figure 1.1). With the exception of hydropower, which is the most mature green technology, it was only in the 2000s that renewables started making significant advances. This applies specifically to solar and wind energy, which account for only 2% and 3.2%, respectively, of global primary energy consumption (Figure 1.1). Still, the pace of renewables deployment (again, especially for wind and solar energy) has been noticeable: global renewable installed capacity (excluding hydropower) has risen from approximately 250,000 MW in 2010 to more than 2 million MW in 2022 (Figure 1.2). Since 2014, the share of net annual additions in global power-generating capacity from renewables has consistently outpaced that from non-renewable energy sources (REN21 2022). Moreover, the contribution of solar and wind energy to global electricity production grew from less than 2% in 2010 to approximately 12% in 2022 (Ritchie, Roser, and Rosado 2023).

The growth in renewables has been driven by the combination of supporting policy measures implemented in several industrialized and developing countries, with an impressive decline in their costs, favoured by the innovation race. The number of countries adopting renewable energy policies increased from a few dozen in the early 2000s to 156 in 2021 (REN 21 2022). In addition, global patents for renewable energy technologies increased fourfold from 2010 to 2021 (IRENA 2022b). In the same time period, the global weighted average levelized cost of electricity (LCOE) of large solar PV projects declined by 88%, while onshore and offshore wind fell by 68% and 60%, respectively (IRENA 2022c). The result is that these green technologies are today competitive with fossil fuel power generation in almost all the regions of the world. However, even if these technologies have become

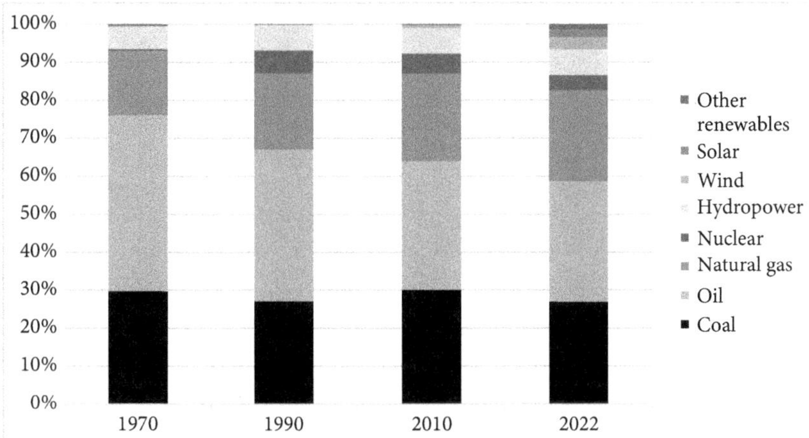

Figure 1.1 World primary energy consumption by source

Sources: Author's elaboration from Ritchie, Roser, and Rosado (2023)

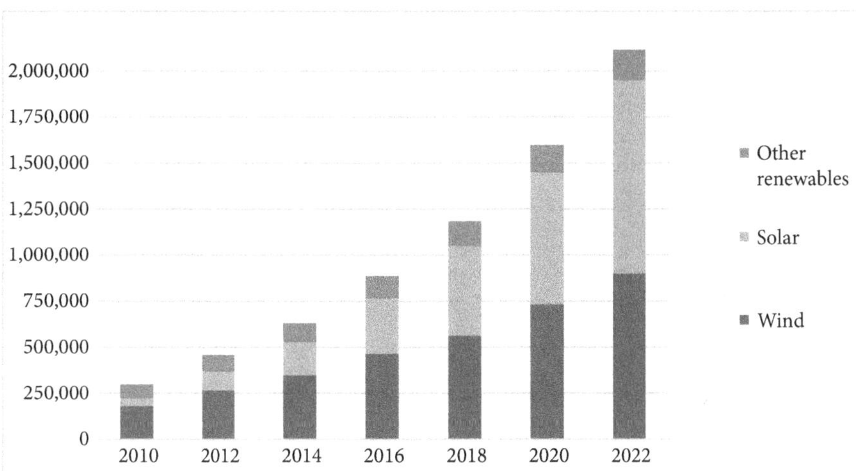

Figure 1.2 Renewable energy (excluding hydropower) world installed capacity (in MW)

Notes: Other renewables include bioenergy, geothermal, and marine energy. Hydropower-installed capacity has increased at a slower pace than solar and wind, from about 1 million MW in 2012 to 1.2 million MW in 2022

Sources: Author's elaboration from IRENA (2018, 2023a)

increasingly cost competitive, appropriate policies continue to remain critically important to break the inertia of a global energy system still based on and designed for fossil fuels. In addition, appropriate investments are crucial in order to speed up the energy transition. These investments have accelerated since the 2008 global financial crisis, first targeting renewable energy

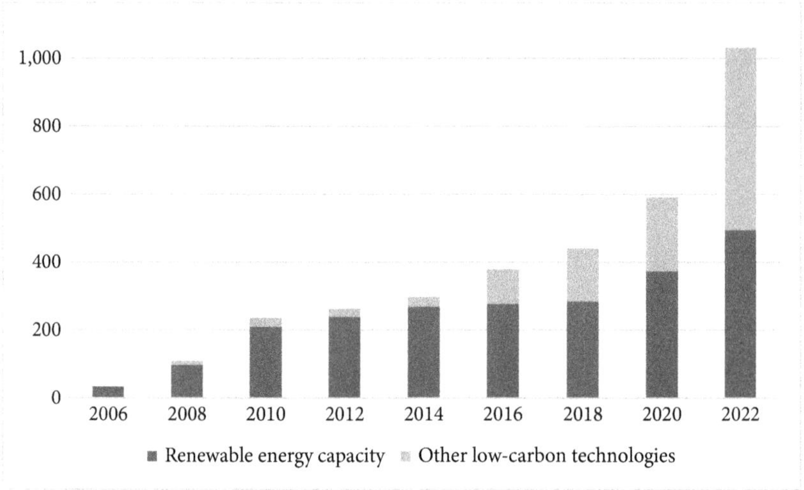

Figure 1.3 Global investments in energy transition (in billion $)

Notes: other low carbon technologies include energy storage, electrified heat, electrified transport (since 2016), and hydrogen (since 2018)
Sources: Author's elaboration from Bloomberg New Energy Finance.

capacity and subsequently other low-carbon technologies, such as battery, EVs, and hydrogen (Figure 1.3).

This trend was confirmed after the Covid-19 pandemic crisis and the Russian invasion of Ukraine: in 2022, global investments in renewable capacity reached a new peak at $495 billion, and overall investment in the energy transition surpassed the $1 trillion threshold for the first time. This result is also significant since the cost of critical minerals and other components of low-carbon technologies has increased in the wake of the Russia–Ukraine war. However, it is worth stressing that subsidies for fossil fuels continue to dwarf the financial resources committed to renewables: in 2020 countries around the world spent almost $6 trillion—approximately 7% of global gross domestic product—on direct and indirect fossil fuel subsidies (REN21 2022). This number further increased in 2021 and 2022, as fossil fuel subsidies became the preferred short-term measure enacted by several governments to mitigate the effects of rising energy prices. Moreover, the gap between green leaders and laggards has widened in the wake of the Covid-19 crisis and the Russia–Ukraine war, exacerbating existing imbalances in the uneven global energy transition landscape (Quitzow et al. 2021; IEA 2022b). Indeed, important regional differences persist in this process. The large part of investments and renewable energy–installed capacity (especially solar and wind energy) of the past decade is concentrated in the Global North, particularly in the EU

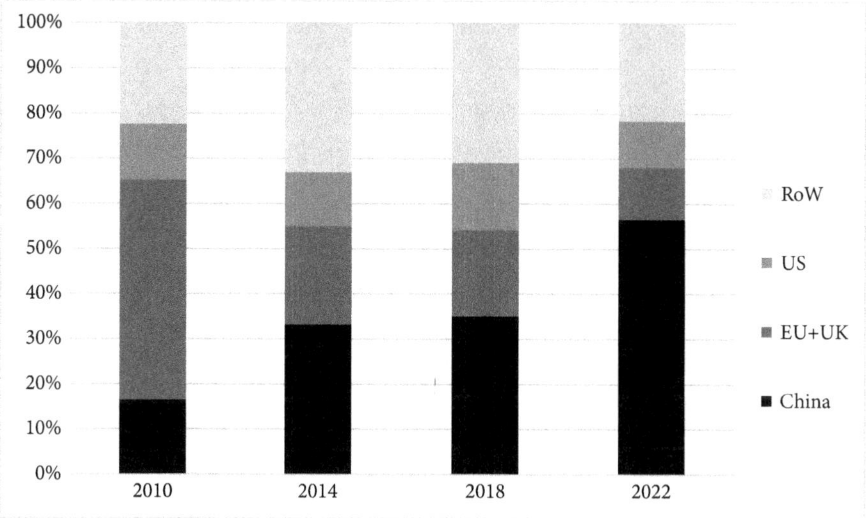

Figure 1.4 Share of global investments in renewable energy

Notes: RoW = Rest of the World; renewable energy = renewable energy power and renewable fuels (e.g., biofuels, green hydrogen). Data for 2010 includes only renewable energy power
Sources: Author's elaboration from REN21 (2021, 2023).

and the US, with the notable exception of China. Beijing became the largest investor in renewable energy technologies during the 2010s, gradually surpassing Europe and taking the lead in these areas (Figure 1.4; Figure 1.5). This trend was confirmed in subsequent years: in 2022 China spent more than $500 billion on energy transition investment (nearly half of the global total), whereas the EU and the US spent only approximately $180 billion and $140 billion, respectively (BloombergNEF 2023). In recent years, China has also steadily increased its role as an innovation centre for low-carbon technologies, particularly in the field of solar energy, although the EU and the US (along with Japan and South Korea) are still prominent actors in the world's green innovation landscape (e.g., IEA 2021d). China, however, currently dominates the manufacturing of key low-carbon technology, from solar PV and wind to batteries and electrolysers (Figure 1.6).

Notwithstanding the advancements achieved in the past decade, existing projections agree that current policies and investment flows will not be sufficient to achieve the Paris Agreement goal of 1.5°C (for a review see IISD 2022; see also IEA 2022b; REN21 2022). However, climate change remains a primary driver of the global energy transition. In 2022, 146 countries—including new large emitters such as China and India—had set net-zero emission targets for which the decarbonization of the national energy system is an

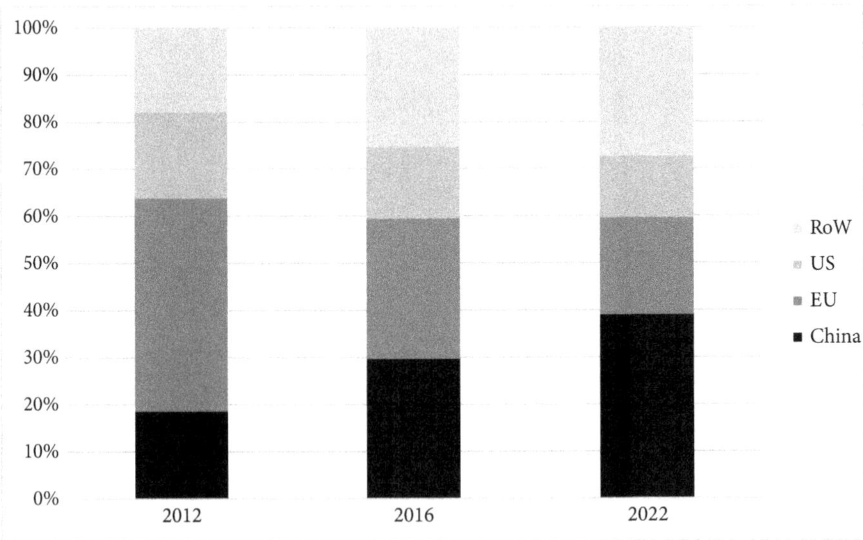

Figure 1.5 Share of solar and wind energy global installed capacity

Note: RoW = Rest of the World
Sources: Author's elaboration from IRENA (2018, 2023a)

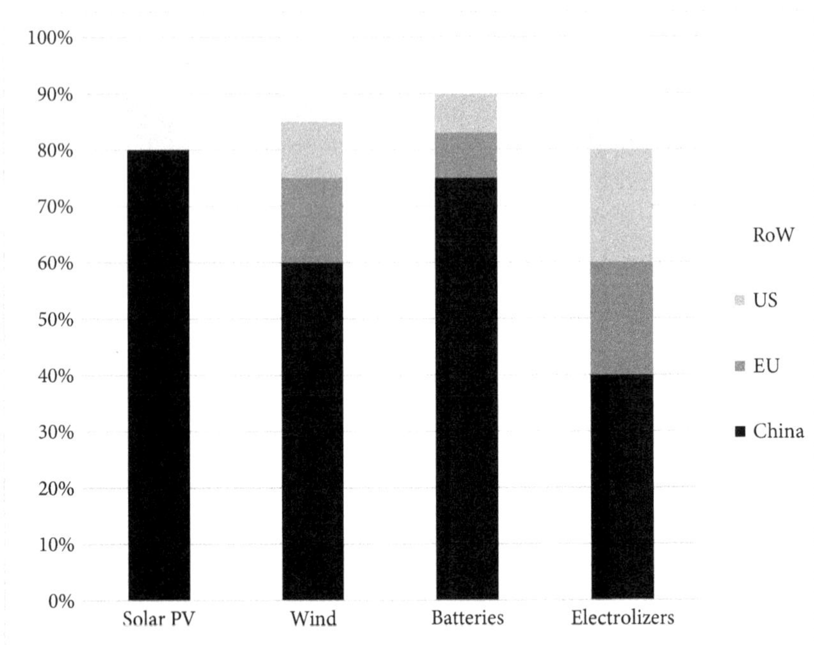

Figure 1.6 Share of global manufacturing capacity for selected green technologies

Notes: Data as of 2022; RoW = rest of the World
Sources: Author's elaboration from IEA (2023b)

essential component (REN21 2023). Environmental concerns reinforced this trend, both in industrialized countries and in emerging economies, where citizens are increasingly concerned about air quality, land and water degradation, and the effects of extreme weather events. A closer integration of energy policy and climate and environmental policy has become the norm for policymakers around the world. The United Nations 2030 Agenda recognized the need to incorporate renewables in its Sustainable Development Goals, with Goal 7 being devoted to affordable and clean energy. This result was the logical evolution of a journey already started with the 1992 UN Framework Convention on Climate Change (UNFCCC), which itself built on the 1987 Brundtland Report, which had highlighted the risks of global warming linked to reliance upon fossil fuels, recommending the development of renewables.

Energy security considerations, in terms of both availability and affordability, were (and are) also important drivers of the diffusion of renewables. After the oil shocks of the 1970s, Western consumers focused on these sources as a means to reduce their energy dependency. With the counter-oil shock and the end of the Cold War, energy security concerns were temporally put aside. However, in the early 2000s, rising oil prices, a structural shift in energy demand towards emerging economies, mainly China and India, and the instability of the international system following the terrorist attacks of 11 September 2001 and the Western military interventions in Central Asia and the Middle East placed this old issue back on the global agenda. The return of energy security concerns was particularly felt in Europe, where energy dependency was growing in parallel with the more assertive foreign policy of Putin's Russia in the post-Soviet space, fuelled by its vast oil and natural gas resources, which were used by Moscow as an energy weapon. Conversely, in the US, the shale revolution enabled the country to significantly increase its oil and natural gas production. Since the late 2000s, Washington's position in global energy markets steadily improved, with the result that the US would soon become the world's largest oil and gas producer. Like in Europe, Beijing's growing dependency on imported oil and natural gas increased the Chinese Communist Party's leaders' interest in renewables. The role of these resources in improving energy security and reducing energy bills once again became central in policy discourse, as fossil fuel prices increased sharply in late 2021 and as a consequence of the 2022 Russian invasion of Ukraine.

Since the 2000s, renewables and low-carbon technologies have become a pillar of a renewed set of industrial policies implemented in industrialized countries and emerging economies alike. After the 2008 global financial crisis, the magnitude and scope of these green industrial policies increased, as many countries included a green component in their recovery strategies.

However, as policymakers in the West realized that China was leading in several low-carbon technologies and related supply chains, these industrial policies gradually acquired a strategic flavour. Leading in the area of renewables and low-carbon technologies was no longer only a matter of assuring adequate and affordable energy to meet national needs while mitigating climate change, but of positioning oneself on the upper end of the low-carbon value chains and at the frontiers of twenty-first-century industrial development. The Covid-19 pandemic crisis subsequently demonstrated, especially in the West, the problems related to the vulnerability of critical supply chains. In addition to oil and gas, Russia is a key global supplier of nickel and other minerals for low-carbon technologies. Consequently, the effects of the Russia–Ukraine war were felt with regard to the security of these supply chains, too. As low-carbon technologies will be the lifeblood of the future decarbonized energy system, the major powers recognized the risks of relying on vulnerable and poorly diversified supply chains, which could be disrupted by unforeseeable events or manipulated by potential rivals. The promotion of renewables and low-carbon technologies was widely included in the recovery plans of major economies post Covid-19. This time, however, the emphasis was more clearly on the contributions of renewables to different sets of problems, such as climate change mitigation, energy security, and economic and geopolitical imperatives linked to the protection (or projection) of national interests. This different framing in relation to renewables, which can be viewed through various lenses, underlines the complexity of the matter, along with the possibility that different actors can emphasize one particular perspective to mobilize resources and increase the legitimacy of the energy transition and steer it in a specific direction. Trade-offs indeed exist among the different objectives subsumed by the different framing of renewables. Geopolitical and national industrial imperatives can be at odds with climate change considerations—for example, if they impede the smooth global diffusion of green technologies or retard their improvement.

A brave new green world order: China, the European Union, and the United States

The transition towards renewables is slower than what is needed to avoid a climate crisis. Fossil fuels are still key ingredients of international affairs. However, a new green world order is emerging: international politics will become increasingly embedded in a global energy and economic system run by renewables, which are increasingly factored in states' foreign policy calculations. This process is unfolding in parallel with the erosion of the Western-dominated liberal order and the forceful return of geoeconomic and geopolitical competition between capitalist great powers (e.g., Buzan and Lawson 2014). Moreover, this competition is of an increasingly economic nature (Blackwill and Harris 2016; Leonard 2016; Wright 2017; Meunier and Nicolaidis 2019; Wigell, Scholvin, and Aaltola 2019; Haroche 2023). In this vein, some scholars have highlighted a 'revival' of economic statecraft and geoeconomics (Wigell, Scholvin, and Aaltola 2019), which can be defined as 'the use of economic instruments to promote and defend national interests and to produce beneficial geopolitical results' (Blackwill and Harris 2016, 20; see also Luttwak 1990). This development is also favoured by the deep economic interconnections between the great capitalist powers, something that was not at work during the Cold War period. In a similar context, asymmetric interdependence has become a 'currency of power' (Leonard 2016, 15) that can be 'weaponized' to achieve foreign policy objectives (Farrell and Newman 2019).

This intertwined relationship between economic means and capabilities and international security issues is amplified by the green transition itself, which is leading to a growing convergence between power and energy influence in international relations (Hafner and Wochner 2020; Bordoff and O'Sullivan 2022; Tocci 2023). In the fossil fuel world, states with no noticeable resources other than oil or natural gas could amplify their posture in the international system by leveraging their energy blessing. Conversely, energy dependency, even for large economies, represented an important constraint on their foreign policy. Although the territorial factor will not disappear in the low-carbon world, the new green energy game is primarily a matter of financial, industrial, and technological capabilities. It is not surprising that the world's three largest economies—i.e., the US, China, and the EU—are emerging as green superpowers by linking their green energy ambition with their global power strategy. Not only are these three actors front runners in the area of green investments, renewables deployment, manufacturing, and innovation, but they can count on other crucial assets for shaping the global energy transition and increasing their influence in the low-carbon world. They can leverage the sheer size of their market to set standards and rules for the nascent green global economy and related international exchanges. As vast markets of truly continental scale and large energy demand, these players can also promote regional connectivity through cross-border infrastructural links, such as transnational power grids (also called 'super-grids', which typically use high-voltage direct current [HVDC] to transmit electricity over long distances) or hydrogen corridors, attracting other states in their sphere of energy influence. Other countries can enhance their position in the low-carbon world by occupying specific niches of the global energy transition. For example, countries such as the Democratic Republic of Congo (DRC), Chile, Australia, and Indonesia are emerging as important players in the area of critical minerals (see Table 1.1). Others, such as South Korea and Japan, have significant industrial and technological capabilities for manufacturing low-carbon technologies. States with vast quantities of cheap solar energy, for example in Latin America or the Gulf, could become important exporters of green hydrogen or related products such as ammonia. Thus far, however, it is only China, the EU, and the US that can mobilize the number of economic, industrial, and technological capabilities that are needed to (at least) compete for the global leadership in the low-carbon world. Moreover, following Buzan and Falkner (2022), there are two other perspectives—i.e., positive and negative power—that highlight why China, the EU, and the US can be considered as green superpowers. As the world's three largest energy and fossil fuel consumers, their inaction could undermine the global energy transition. In addition, as demonstrated, they control important green 'goods', which can promote this process, such as financial, technological, and industrial capabilities. In both cases, their significant negative power assigns them a de facto veto power in international green affairs, as they can slow down or weaken global efforts in this area. Positive power—that is, the ability to engender a positive change in international politics—is closely related to the concept of international leadership, which is based on the notion that leaders are needed to propose cooperative solutions and innovative policy ideas, provide technological and economic aid in support of certain policy objectives, and push for international consensus on challenging issues (Buzan and Falkner 2022). Such leadership can be based on intangible assets, such as ideas, policies, and institutions, or on the use of material assets and incentives, such as finance, to change actors' behaviour (Young 1991). The economic might, political power, and diplomatic clout of China, the EU, and the US assign them a systemic importance across most, if not all, areas crucial for pushing forward and shaping the global energy transition, including industry, climate change, and trade and investment regimes. In summary, the combination of the negative and positive power resources of China, the EU, and the US is still not matched even by large emerging economies such as India or Brazil. However, this situation could change in the near future.

The fact that the new green energy game is played primarily in the economic realm explains the focus on the EU. The EU is traditionally viewed as a military dwarf but an economic superpower with a global regulatory reach (e.g., Bach and Newman 2007; Bradford 2012, 2020; Damro 2012; Lavenex,

Krizic, and Serrano 2017). The EU institutions, and especially the EU's executive body, the European Commission, regulate the single European market—which is larger than the US market—and have key responsibilities over the bloc's external economic and trade-related policies. In the field of energy, the EU has widely increased its competences in recent decades, particularly after the Lisbon Treaty came into force in 2009. The EU member states still maintain important powers over national energy issues, but their decisions are taken within the framework of the EU energy and climate policy. Moreover, the EU has developed a significant normative and ideational stance in international affairs (e.g., Manners 2013). Its efforts to play a leadership role are particularly evident in the areas of global environmental politics (e.g., Biedenkopf, Dupont, and Torney 2022), climate change (e.g., Wurzel and Connelly 2010; Wurzel, Connelly, and Liefferink 2016), and, recently, energy transition (e.g., Siddi 2023; Tocci 2023). To be sure, long-standing questions persist about the EU's actorness and coherence as a major power in global politics (Bretherton and Vogler 2005; Vogler and Bretherton 2006). In addition, several differences obviously remain between the EU and traditional national state actors such as China and the US. These issues are discussed further on in the book. However, the EU can be analysed along with these other green superpowers.

Important differences exist in the (green) political economies of China, the US, and the EU as well. In Europe and the US, the diffusion of renewables and low-carbon technologies occurred in the context of liberalized, privatized energy markets. In both cases, especially in Washington, a market-oriented approach has long informed energy and climate policy. This approach does not mean that state interventionism has been minimal nor that it has played a negligible role. However, China's green trajectory has, since the beginning, been embedded in Beijing's state-led (and party-led) model of capitalism, based on the tight control of markets by the government and the Chinese Communist Party and on a prominent role for state capital in energy and economic governance. Recently, China has gradually shifted towards more market-based tools in the area of renewable energy and climate change; however, these features still represent a significant and defining aspect of Beijing's green development and leadership. In this regard, it is also important to reiterate the better positioning of China in several segments of the low-carbon technology supply chains; from upstream (mining and processing of critical minerals and REEs), to midstream (manufacturing) and downstream (deployment). China's prominence is particularly evident in the manufacturing of key green technologies. The country is home to the world's ten top suppliers of solar PV equipment. Moreover, China's world share in the manufacturing of solar PV technologies (particularly polysilicon and solar panels) is almost 80%, which is more than double its share of global PV demand (IEA 2023a, 2023b). Conversely, the EU imports nearly all its solar panels, while the US covers approximately two-thirds of its demand by imports (IEA 2023a). The wind business is less concentrated, with an important role to be played by European and US companies (the EU is a net exporter of wind turbines, while the US is a net importer). Nevertheless, China accounts for approximately 60% of global wind manufacturing capacity and it is home to six of the world's top ten wind turbine companies (GWEC 2022). China also dominates the manufacturing of electrolysers for hydrogen (although European companies lead the field of innovation for green hydrogen technology) and batteries (lithium-ion batteries). Batteries, in particular, account for 30%–40% of the value of EVs, for which the country is the world's top market ahead of the US and the EU (China is also ahead in the construction of new battery gigafactories).

In addition, China dominates the processing of many of the critical raw materials required to deliver energy transition, namely cobalt, copper, nickel, lithium, and (especially) REEs (Table 1.1.). In the case of REEs, Beijing's dominance extends to the mining sector as well: 60% of the world's REE production is concentrated in China (Table 1.1). China also holds a large part of the world's REE reserves, as well as important reserves of other critical minerals (Table 1.1). Moreover, Chinese companies are involved in numerous mining activities abroad. For example, they own the majority of the cobalt mines of the DRC and they have substantial investments in lithium production in Australia and Latin America and nickel in Indonesia.

On the other hand, both the EU and the US are primarily importers of critical minerals and REEs. The EU, in particular, does not have significant reserves of critical minerals, although certain Eastern European countries such as Poland, Czech Republic, and Serbia do have some copper and lithium deposits. However, proven reserves of most critical minerals are more geographically widespread than current production, suggesting that there is considerable scope for diversifying the sources of supply for consumers (Table 1.1). Several actual and potential mineral producers are also less problematic, especially for Western countries, than many oil and gas exporters; they include countries in politically stable regions (such as Latin and Central America), democratic allies such as Australia and Canada, and possible energy partners such as Chile, India, Brazil, and Indonesia (Table 1.1.). Finally, rising demand can foster exploration activities and hence lead

Table 1.1 Major proven reserves' holders, and mining and processing countries for selected critical minerals and REEs (global percentage)

Cobalt	Copper	Lithium	Nickel	REEs
Proven reserves (*)				
DRC (46%)	Chile (22.7%)	Chile (41.8%)	Indonesia (22%)	China (35%)
Australia (18%)	Australia (10.6%)	Australia (26%)	Australia (22%)	Vietnam (17.7%)
Cuba (6.6%)	Peru (8.8%)	Argentina (10%)	Brazil (16.8%)	Russia (16%)
Russia (3.3%)	Russia (7%)	China (6.8%)	Russia (7.9%)	Brazil (16%)
Canada (2.2%)	Mexico (6%)		Philippines (5%)	India (5.5%)
	US (5.5%)		China (2.9%)	Australia (3.2%)
	DRC (3.5%)			US (1.5%)
	Poland (3.5%)			
	China (3%)			
Mining (**)				
DRC (70%)	Chile (23.6%)	Australia (46.9%)	Indonesia (48.8%)	China (60%)
Indonesia (5.4%)	Peru (10%)			US (13%)
Russia (4.8%)	DRC (10%)	Chile (30%)	Philippines (10.1%)	Myanmar (11%)
Australia (3.2%)	China (8.6%)	China (14.6%)	Russia (6.7%)	
Canada (2.1%)	US (5.9%)	Argentina (4.7%)	France (New Caledonia) (5.8)	
Cuba (2%)	Russia (4.5%)	Brazil (1.6%)		
	Indonesia (4.1%)		Australia (4.9%)	
	Australia (3.7%)		Canada (4%)	
Philippines (2)	Zambia (3.5%)		China (3.3%)	
	Mexico (3.3%)		Brazil (2.5)	
Processing (**)				
China (70%)	China (42.3%)	China (58%)	Indonesia (39.8%)	China (88%)(*)
Indonesia (18%)	Chile (8.1%)	Chile (29%)		Malaysia (8%)
Finland (11%)	DRC (6.5%)	Argentina (10%)	China (23.9%)	
	Japan (6.2%)		Japan (5%)	
	US (3.9%)		Canada (3.7%)	
	Russia (4.2%)		Australia (3.6%)	
	South Korea (2.5%)			
	Germany (2.4%)			

Notes: (*) = as of 2020; (**) = as of 2023
Sources: Author's elaboration from IEA (2021c); Leruth et al. (2022); and IRENA (2023b)

to the discovery of new reserves. For example, the mining potential of several African countries is still largely unknown (Wouters 2023). Creating new supply chains, however, remains a challenging process; mining is a long lead-time, capital-intensive sector with (often) high environmental costs for local communities and high risks in terms of conflicts and human rights abuses (e.g., Church and Crawford 2020). Processing as well has a considerable environmental impact. Moreover, resource nationalism, which has increased in recent years, could negatively affect global supply and slow down investments (IRENA 2023b).

China's industrial, processing, and mining capacities—and lower concerns for the negative environmental impacts of these activities—have been instrumental in bringing down the costs of low-carbon technologies, helping their global diffusion. At the same time, China's prominence has triggered the competitive response of the US and Europe, which are concerned about their vulnerability and dependency in low-carbon technology supply chains; they are also concerned about losing the green race and becoming technology takers rather than makers in key sectors of the twenty-first-century global economy. These concerns increased after the launch of China's Belt and Road Initiative (BRI), which has gradually incorporated a green component, mirroring abroad the Chinese green state-led developmental approach and expanding the external sphere of Beijing's influence. In the aftermath of the 2008 global financial crisis and the 2010 China–Japan REE dispute, the Obama administration was the first to arrange a response to Beijing's green leadership. Later on, although with important differences, both the Trump and the Biden presidencies improved these efforts. Under Trump, Washington focused primarily on exploiting the results of the shale revolution. The Biden administration, however, has not only reengaged the country with global climate change policy but has also brought about a sharper focus on relaunching green investments and manufacturing at home—notably with the Inflation Reduction Act—and promoting diversification abroad, prioritizing friends and allies. In the EU, too, green industrial policy accelerated after the 2008 global financial crisis under the Barroso Commission. However, it was later, under the Juncker Commission—and more clearly under the von der Leyen Commission in the wake of the Covid-19 crisis—that the EU developed a more comprehensive and strategic approach to the energy transition. The 2022 war in Ukraine—amid a deterioration in relations between China and the West—has further accentuated the foreign policy and strategic implications of the global energy transition.

Structure of the book

The goal of this book is to analyse, compare, and explain the emerging green foreign energy policies and green power strategies of China, the EU, and the US. Fossil fuels have been key to major powers' foreign policies for a long time. In the context of the current global energy transition, renewables and low-carbon technologies are emerging as elements that can have a similarly important impact on twenty-first-century world politics. In Chapter 2, I present and discuss the analytical framework of the book. First, in this chapter I highlight the main differences between the traditional fossil fuel foreign energy policy and the green one for the low-carbon world. I offer a definition of green foreign energy policy and I illustrate its building blocks. Next, I discuss in more detail the main drivers of green foreign energy policy, the issue areas (i.e., climate, energy, industry, trade and investments) in which it can be analytically organized and empirically mapped and the statecraft instruments (i.e., diplomatic, economic, and regulatory) at the disposal of governments to achieve their objectives in this field. I go on to elucidate the framework for analysing and comparing major powers' green foreign energy policies. This framework is based on a four-dimensional matrix that comprises an issue area dimension, a tools dimensions, a mode of engagement dimension (incorporating the unilateral–multilateral spectrum), and a geographical dimension. Empirical examples are provided in this chapter to clarify how this four-dimensional matrix can help map green foreign energy policies and trace their variations among cases and over time. In this chapter, I also explore the internal–external nexus in green foreign energy policy. I argue that this nexus sheds light on how the evolving patterns of state–market relations at the domestic level in the area of renewables and low-carbon technologies influence major powers' green external action. Lastly, in Chapter 2 I illustrate the various green power strategies that major international actors can pursue in the wake of the global energy transition. These green power strategies are related to the capabilities at their disposal, their ambitions, and their strategic choices regarding the mix of tools to be deployed. These choices are influenced by state–market relations and by the opportunities and constraints posed by the internal and external environments. I focus on different conceptualizations of power to characterize major international actors' green power strategies: soft power, regulatory power, economic power, and catalytic power. These conceptualizations are adapted for understanding the approaches to the global energy transition being followed by major international actors, both from a comparative perspective and diachronically, in order to make sense of the evolution of green power strategies over time. In particular, a green soft power strategy relies on intangible assets and means in terms of ideas and polices and aims at shaping the milieu in which the global energy transition takes place. Economic green power illustrates a strategic approach with a mercantilist edge. The regulatory and catalytic perspectives, on the other hand, are closer to a liberal approach to green energy relations, although the latter perspective relies on different tools for mobilizing resources and achieving foreign policy objectives.

Chapters 3, 4, and 5 constitute the empirical core of the book, in which I illustrate the origins, drivers, and modalities of the green foreign energy policies of the green superpowers: China, the EU, and the US. In these chapters I present further data on their energy situation and for each case I discuss significant developments in terms of state–market relations, internal–external nexus, and green power strategies. The main empirical focus of these chapters is on the period spanning from the early 2010s (after the 2008 global financial crisis) to 2022–2023, a time characterized by the post-Covid-19 recovery and the international effects of the Russian invasion of Ukraine. In the case of China, the focus is on the years covered by the 12th, 13th and 14th Five-Year Plans (2011–2015, 2016–2020, and 2021 to date). In Chapter 3, I also analyse the green foreign energy policy dimension of the BRI. In the case of the EU, the focus is on the period of the second Barroso Commission (2010–2014), the Juncker Commission (2014–2019), and the von der Leyen Commission (2019–2024). Finally, in the case of the US, the analysis covers the Obama era (2009–2016), the Trump presidency (2017–2020), and the Biden presidency (2021–2024), which relaunched Washington's internal and external green energy policy. As for the sectors, the empirical chapters cover the most promising renewables and low-carbon technologies discussed to date—that is, solar and wind energy, batteries, EVs, electricity grids, and green hydrogen—and their supply chains, including those critical minerals and REEs essential for their manufacture. It is important to stress that several critical minerals are also crucial for the high-tech and defence industry. However, I focus only on their implications for the energy transition. Moreover, as this book focuses on the green renewables component of the current transformation, I do not cover other technologies that can play a role in the decarbonization of the global energy system, such as nuclear energy, carbon capture, utilization, and storage (CCUS), or carbon removal technologies; although all these technologies are also part of the Chinese, European, and US efforts to reduce fossil fuel consumption and/or CO_2 emissions. Nor do I discuss transversal issues that can impact the functioning of the nascent green energy system, such as cybersecurity, which is a challenge linked to the digitalization of several economic sectors rather than a specific risk for the energy transition itself. Finally, in this book, I do not consider the international role of subnational actors, such as cities. While this is an important theme for the global energy transition, it is nevertheless largely covered by the literature on transnational climate governance and related studies (e.g., Bulkeley at al. 2014).

After the empirical chapters devoted to each case study, in Chapter 6 I present a structured, focused comparison of the three green superpowers based on the analytical framework developed in the second chapter. In this chapter, I first offer a comparative assessment of the historical trajectories of their green foreign energy policies and green power strategies by highlighting the main turning points and developments in recent decades. Then, I illustrate and explain in detail the differences and similarities among China, the EU, and the US. I also problematize the study of the EU as an international actor and a major power by referring to insights from EU studies. Although the three green superpowers demonstrate certain commonalities, their diverse political economies and patterns of state–market relations translate into specific external green approaches. Chinese green state-led capitalism is mirrored in the 'going-out' green strategy of the country, while the US's more market-oriented approach—which, again, does not mean a marginal role for state agencies and policies—is reflected in the country's green foreign energy policy. The same applies to the EU, which has gradually acquired a catalytic role in the area of renewables and low-carbon technologies, both within Europe and in its external action. Chapter 6 concludes with an assessment of the effectiveness of the Chinese, EU, and US green power strategies and a discussion of their limitations and potentials. These limitations and potentials consider both internal capabilities and weaknesses and the constraints and opportunities offered by the changing international landscape at the regional and global levels. China's green power strategy, which is mainly based on economic power and close coordination of the state-industrial complex, has been effective in supporting the country's rise to the top of the green technology ladder. Not only is China a technology maker in many relevant sectors of the global energy transition, but it has also extended the scope and geographical reach of its external influence thanks to its achievements in the area of renewables and low-carbon technologies. However, this success has triggered counter-reactions that have gradually worsened the international environment for China's external action. Beijing's efforts to forge strategic partnerships and extend its green economic power will be increasingly constrained, while other major powers will try to reduce their exposure to Chinese-led low-carbon technology supply chains, challenging the very basis of Chinese influence. At the same time, China has recently tried to advance soft and regulatory green power strategies. The latter, however, requires capabilities that are still underdeveloped in the Chinese context. Chinese green soft power as well, although attractive for some countries in the Global South, has to face domestic constraints. The evolution of the EU's green power strategies in recent decades entails a gradual shift from soft power to regulatory power to (more recent) efforts towards building catalytic power capabilities and strategically combining the different instruments of the EU's toolbox in (primarily) mini- and multilateral modes of engagement. Although the EU has developed a flexible toolkit for green foreign energy policy, the effectiveness and geographical reach of its action is limited by strategic incoherencies, an internal capabilities gap, and a more competitive international landscape that represents an obstacle to its traditional liberal approach. Like in the EU, the green power strategy of the US has evolved from soft power towards the deployment of additional resources and tools to increase its external influence in the emerging low-carbon world. However, only recently, the US has widened and sharpened its armamentarium to play a more effective role in the global energy transition, both domestically and abroad. Its green power strategy has intensified the use of economic power with growing mercantilist elements and is increasingly linked to the country's military posture as new alliances and strategic partnerships are formed (or strengthened) with a green energy component attached to them. The hegemonic status of the US—although in (relative) decline—along with renewed domestic efforts can help the country regain a leadership position in the global energy transition. However, domestic constraints, the politicization of climate change, and a declining role in some regional theatres can hinder the green power strategy of the US.

Chapter 7 concludes this book by briefly reassessing its main empirical findings and contributions and discussing the wider implications of Chinese, EU, and US policies and strategies for the emerging green world order. In addition, in this chapter I consider insights into possible future scenarios on the matter. I argue that competition between green superpowers will become a new normal in the twenty-first century. The key will be to find an appropriate balance between competitive tensions and the level of cooperation and economic interdependence required to speed up the global energy transition and meet climate mitigation targets. Finding this balance will be very complex, as many uncertainties remain regarding the long-term political and economic effects of the global crisis triggered by the Covid-19 and the Russian invasion of Ukraine. Climate change efforts can help find this balance. In this case as well, however, uncertainties remain regarding the real commitment of major powers to decarbonization, which will be more and more affected by domestic politics, as the structural transformations generated by the energy transition will increase resistance to serious climate action. The rivalry between China and the US will be the driving force of the competitive dynamics related to the global energy transition. In the US, politicization over climate change will represent another important challenge, whereas in China it will be difficult to overcome fossil fuel lock-in that risks undermining Beijing's green leadership. However, these countries' efforts to increase their influence in the low-carbon world—including the creation of new green alliances—can have positive spillovers for decarbonization and climate change. The EU will navigate between different, sometimes contrasting, incentives. Its plans to achieve greater autonomy over strategic low-carbon technologies will be implemented and its efforts to develop more coherent and strategic external action on the matter will be confirmed. However, cooperation with China and the US will be preserved and even improved in some sectors to tap capability gaps and to achieve wider foreign policy objectives. This could create problems in transatlantic relations if China–US rivalry escalates further. However, the EU could also emerge as a 'fixer' that can alleviate the competitive tensions emerging from the global energy transition, as its weaknesses can turn into opportunities for multilateralism and preserving integration along renewables and low-carbon technology supply chains.

2

Green Foreign Energy Policy

A Framework for the Analysis

Energy and foreign policy: from fossil fuels to renewables and low-carbon technologies

Since the famous decision of Winston Churchill, as First Lord of the Admiralty, to power the British navy's ships with oil instead of coal, energy has been deeply intertwined with foreign policy (Yergin 2006). This decision implied that the national security and international posture of the hegemonic power of that time was no longer reliant on domestic sources—coal was extracted mainly in Wales—but on external supplies from distant territories, particularly what was then Persia, across several sea lines. These new vulnerabilities would soon become the focus of major powers' foreign policy as the First and Second World Wars highlighted the critical role of oil in modern warfare. The notion of energy security emerged and was institutionalized in the discourses and practices of foreign policy in this period (e.g., Cherp and Jewell 2011; Dannreuther 2017). This term was used synonymously with 'security of supply' as several countries, in both Western Europe and in Asia, such as Japan, found themselves in the difficult position of being consumers and importers of oil. Despite the fact that the new powers that emerged from the devastation of the Second World War, the US and the Soviet Union, had abundant fossil fuel resources, energy security—framed as a risky dependence on supplies from abroad—became one of the top priorities for decision makers in industrialized countries. This occurred in parallel with the penetration of oil into their economies. The events that brought about the first and second oil shocks, in the 1970s, have done nothing but reaffirm—and amplify—the nexus between fossil fuels and foreign policy.

One of the outcomes of those energy crises was the development of the institutional energy security landscape. The US promoted the creation, in 1974, of the International Energy Agency (IEA) to counter Organization of the Petroleum Exporting Countries (OPEC, established earlier, in 1960) and increase policy coordination among consumers members of the

Green Superpowers. Andrea Prontera, Oxford University Press. © Andrea Prontera (2024).
DOI: 10.1093/9780191987304.003.0002

Organisation for Economic Co-operation and Development (OECD). Both the IEA and OPEC have evolved over the time, increasing their membership and scope. Also, they have begun to cooperate on certain specific questions, such as information sharing and data exchanges. However, the structural diversity of interests between consumers and producers has inhibited the creation of common global institutions for managing international energy relations. After the end of the Cold War, Western countries attempted to include energy into new global (World Trade Organization (WTO)) and regional (Energy Charter Treaty and North American Free Trade Agreement (NAFTA)) trade regimes. The results of these efforts have been limited, touching only the surface of international energy relations. The global institutional landscape for fossil fuel foreign energy policy has remained organized around a system of competing 'clubs', which are composed of producer and consumer countries.

Another important consequence of the 1970s oil shocks was the accelerated diffusion of natural gas. This new fossil fuel began to be a factor in the discourses and practices of foreign energy policy. Unlike oil, which became a globally traded commodity, natural gas exchanges developed around regional markets (Eurasian, North American, East Asian, and Latin American). Since the early 2000s, with the expansion of the liquified natural gas trade, connections among these regional markets have increased. However, this development has not yet eliminated the differences between the international political economy of oil and that of natural gas, with the latter being linked to regional dynamics and pipeline politics (MIT 2011; Hulbert and Goldthau 2013). This also explains why the efforts to create a 'club' of gas-exporting countries—as with the Gas Exporting Countries Forum—have not reached the same level of institutionalization as OPEC.

In the fossil fuel world, oil and natural gas have become an end and a means of major powers' foreign policies. In both cases, however, the main issues have revolved around two intertwined key elements: territoriality and flows. This combination results from the geographical concentration of fossil fuel resources, which makes some regions more valuable than others, and the consequent internationalization of energy trade. The manipulation of flows of energy resources—i.e., the 'weaponization' of oil and gas—to achieve non-energy-related foreign policy objectives follows the traditional logic of economic statecraft and can be engaged in by both producer and consumer countries (e.g., Newnham 2011; Fischhendler, Herman, and Maoz 2017; Van de Graaf and Colgan 2017). Oil and gas can be used qua positive and negative economic sanctions to punish (e.g., embargos or boycotts) or reward (oil and gas being supplied at 'discounted' prices) other states, win friends, push a national foreign policy agenda, or promote a policy or regime change. Conversely, the key strategic goals of foreign energy policy in the fossil fuel word, at least from the perspective of consumer countries, revolved around ensuring the uninterrupted availability of oil and natural gas and promoting diversification of suppliers and supply routes, to reduce vulnerability to supply disruptions or the political leverage of producers. These goals implied several related operational objectives, such as promoting political stability in the major producing regions, supporting the smooth functioning of the international energy markets, preventing and managing energy crises, promoting the development of energy infrastructures, and patrolling strategic sea-lines of communication and choke points that allow for the free flow of resources and practical matching between supply and demand. In attempting to achieving these objectives, states have traditionally used a combination of all the foreign policy tools at their disposal. Military force (its use or the threat of using it) can be deployed to control territories rich in fossil fuel resources or patrol strategic sea lines of communication and choke points in international energy markets, such as the Middle East straits. Since the Second World War, however, only the US has had at its disposal those capabilities required to develop a foreign energy policy supported by a consistent military hedge (e.g., Stokes and Raphael 2010; O'Sullivan 2013). The other consumers have mainly relied on different economic tools and modes of diplomatic engagement to convince producer states to exploit their energy resources, open up their markets to foreign energy companies and investments, and act as reliable suppliers. This is also the case for China. Despite the fact that its naval posture has been strengthened in the last decade, Beijing has not yet developed those capabilities for military force projection that can support a muscular foreign energy policy in the more important of the world's producing regions. Indeed, China has relied on a combination of economic assistance and bilateral diplomatic engagement, coordinated with its state-owned energy companies, to develop its foreign energy policy towards producer countries in the Middle East, Africa, and Central Asia (e.g., Kong 2009).

The territory–flows combination has been matched by another important pair in the foreign energy policy of the fossil fuel word, that is, government–company interactions (e.g., Anderson 1981; Keohane 1982; Randall 2005; Abdelal 2013; Högselius 2013; McFarland and Colgan 2023). International oil and gas markets are dominated by a small number of companies, the so-called oil and gas majors. Obviously, there are many other actors involved in foreign energy policy formulation and implementation. However, governments and oil and gas majors are the main players in the fossil fuel energy game, and their (dynamic) interactions are key to understanding states' strategies on the matter (e.g., Prontera 2017, 2019). In this sense, the policy-making process of the foreign energy policy in the fossil fuel word resembles Lowi's elitist model (Lowi 1967), with few actors involved, particularly when energy security considerations of national significance are at stake. In this latter case, the heads of executive and the foreign policy ministries are directly involved.

Renewables and low-carbon technologies can also be seen as foreign policy ends (e.g., diplomatic and financial support for promoting their deployment abroad or favouring the internationalization of national green industries) and means (e.g., leveraging green trade relations to increase influence over other regions/countries or manipulating low-carbon technologies trade to punish a target country). The ongoing shift to renewables, however, has further increased the complexity of the relationship between energy and foreign policy. Firstly, military tools, which have played an important role in the fossil fuel era, cannot be used to extend control over the sun or the wind. Unlike fossil fuels, renewables are not geographically concentrated in a handful of regions. Moreover, renewables per se are not traded internationally. As explained, the new green energy game is mainly a matter of technological innovation and industrial partnerships, rather than patrolling international oil market choke points or extending control over fossil fuel–rich territories. Indeed, the shift towards renewables has added a new crucial element to the territory–flows combination, i.e., the transnational space of the low-carbon technology global supply chains. Military tools have limited value in increasing influence over these supply chains. Under certain conditions, military tools will still be important in the low-carbon world, but only as an element of wider green power strategies. To be sure, territoriality and flows have not disappeared as important factors in the emerging green energy game. Those regions rich in the critical minerals and REEs required for renewable and low-carbon technologies may be the object of new competitive dynamics among countries. REEs trade flows can be 'weaponized' to achieve non-energy-related foreign policy objectives, as illustrated by the China's embargo of REEs exports to Japan in 2010. Similarly, cross-border electricity trade and green hydrogen trade can create new dependencies that may be exploited by states in the 'energy weapon' fashion (e.g., Blondeel et al. 2021). However, as illustrated in Chapter 1, these elements are not sufficient to reproduce the kind of geopolitical confrontational scenarios seen in the fossil fuel word.

Tools of economic statecraft and diplomacy will be more relevant to the nascent green foreign energy policy. In this field, governments must navigate the intricate transnational space of low-carbon technology supply chains.

They must reduce their vulnerabilities in these supply chains while, at the same time, favouring those levels of innovation and costs reduction that are required to be at the front of the global green technological race and meet challenging climate targets. Moreover, given the complexity of and interdependencies within energy systems, the deployment of renewables is contingent on the implementation of corresponding institutional (i.e., regulations and standards) and social changes (i.e., behavioural changes among users and other stakeholders) (Quitzow and Thielges 2022). The global energy transition is an open-ended process that entails a reconfiguration of the entire socio-technical energy system (Van de Graaf and Sovacool 2020). Shaping the form of this system—by promoting the preferred rules, institutional settings, and views—can assign states a leadership role in the emerging low-carbon world. Socio-technical energy systems are imbued with normative elements, because of the values and principles that underpin them, which (for instance) can be more or less just, more or less sustainable, more or less democratic and transparent (e.g., Van de Graaf and Sovacool 2020; Szulecki and Overland 2020). Shaping the global energy transition, hence, also means to support certain values and principles orienting this transformative process and its outcomes.

Overall, because renewables and low-carbon technologies are, at the same time, at the core of the energy–climate nexus and a key ingredient in twenty-first-century industrial development, green foreign energy policy's strategic goals revolve around governments' external actions that aim to promote and shape the transition from fossil fuels to renewables, reducing vulnerabilities along low-carbon supply chains while supporting national green business. External actions to promoting the energy transition include those measures sustaining the diffusion of renewables abroad, as well as those supporting the energy transition at home that target third states, such as the realization of cross-border electricity networks or green hydrogen corridors for the import of renewable energy. However, from an analytical perspective, it is important to stress that the increasing relevance of supply chains blurs the distinction between domestic and external measures. Reducing vulnerabilities along low-carbon supply chains can include supporting industrial capacity, green manufacturing, or mining at home, as well creating stockpiles of critical minerals or promoting battery recycling. But these efforts can be combined with actions abroad to (re)direct some supply chains' segments towards allies or political aligned countries; a phenomenon dubbed as 'friend-shoring'. In other words, in the context of the global energy transition, managing low-carbon technology supply chains is a constitutive element of national green foreign energy policy (e.g., O'Neil 2021). Moreover, unlike in the fossil fuel word, where the interactions between governments and oil and gas majors have monopolized the field, the emerging low-carbon world is characterized by a multiplication of state and non-state actors. This situation challenges the previous elitist model of foreign energy policy, opening the field up to more pluralistic policymaking processes. These involve several branches of executive and governmental agencies—because renewables and low-carbon technology development cuts across numerous sectors—and a plurality of non-state actors, including a variety of companies and business organizations working in different segments of green industries and supply chains. An additional challenge to the previous elitist model is presented by the energy–climate nexus. This nexus has further expanded the number and variety of actors involved in green foreign energy policy from non-governmental organizations (NGOs) to international financial institutions (IFIs) and development agencies from the donor community. Finally, the context in which green foreign energy policies are formulated and implemented is further complicated by the institutional landscape that characterizes the global energy transition. The global governance of renewables arose by default, rather than design, through the ad hoc interplay of various aspects of the international economic and political system (Leal-Arcas and Minas 2016; see also Hirschl 2009; Lesage, Van de Graaf and Westphal 2010; Chelminski, Andonova and Sun 2022). Several institutions involving different actors address different aspects of renewables development. Some of the components of this fragmented system are global in scope, such the WTO and the IRENA. In others cases, renewables have been partially embedded in regional agreements and organizations, such as NAFTA (now USMCA) or the ASEAN. Global forums, such as the G7 and G20, also have increased their focus on renewables and low-carbon technologies since the late 2000s, in parallel with a growing number of bilateral cooperation agreements, efforts of regional and multilateral development banks, or public–private initiatives. Moreover, because of the climate–energy nexus, renewables are closely linked to the 'regime complex' for climate change (Keohane and Victor 2011). This adds further complication to the institutional landscape of green foreign energy policy, which is better described precisely by the notion of 'regime complex', an array of partially overlapping, fragmented, and non-hierarchical institutions (e.g., Chelminski, Andonova, and Sun 2022).

Table 2.1 summarizes the key differences between the traditional foreign energy policy in the fossil fuel word and the emerging green foreign energy policy in the context of the global energy transition. Within this transformative process, states must navigate sometimes contrasting incentives and trade-offs. Reducing vulnerabilities along low-carbon supply chains

Table 2.1 Fossil fuel foreign energy policy versus green foreign energy policy

	Foreign energy policy **Fossil fuel**	**Green**
Main focus	Territory Flows	Territory (-) Flows (-) Supply chains (+)
Main goals	Ensuring the uninterrupted availability of oil and natural gas	Promoting and shaping the transition from fossil fuels to renewables
	Promoting the diversification of suppliers and supply routes	Reducing vulnerabilities along low-carbon supply chains Supporting national green business
Main tools	Military Economic Diplomatic	Military (-) Economic (+) Diplomatic (+)
Main actors	Governments (e.g., heads of the executive, foreign policy ministries)	Governments (several branches of the government)
	Companies (a few oil and gas majors)	Companies (several companies active in different segments of green business) Business organizations, NGOs, IFIs
Policymaking process	Elitist	Pluralist
Institutional landscape	Opposing 'clubs'	Regime complexes

Notes: (-) = decreasing relevance; (+) increasing relevance
Source: Author's elaboration

and supporting national green business can result in protectionist measures that can hinder the global energy transition and the achievement of climate targets (e.g., Goldthau and Hughes 2020).

Drivers, issue areas, tools, and modes of green foreign energy policy

The differences between the traditional foreign energy policy and the green one concern additional elements beyond those sketched above. The main driver of foreign energy policy in the fossil fuel word was addressing energy security concerns. Conversely, a combination of factors lies behind the growing government intervention in the emerging field of green foreign energy policy. In the 1970s and 1980s, when energy security was a hot topic for major consumers, promoting renewables gradually became a goal of their international development policies. In several producing regions, such as in the Middle East and North Africa, US, European, and Japanese development agencies began to support research and pilot projects on renewables, mainly solar energy, which has enormous potential in those regions. This move had two objectives: to help these countries meet their energy needs and to free additional oil and gas resources for export. However, these initiatives were still marginal and did not represent a manifestation of an explicit green foreign energy policy; for the very reason that renewable energy policies were yet underdeveloped at the domestic level. In the late 1980s and 1990s, climate change emerged as a main driver of the nascent major powers green foreign energy policies, which were oriented towards the multilateral efforts that resulted in the 1992 UNFCCC and later the Kyoto Protocol. Since this period, renewables also began to enter more firmly into the agenda of the donor community and in the package of development policies implemented by UN agencies. Now, the promotion of renewables abroad was intended as a means of contributing to climate and sustainable development objectives, along with helping meet energy demands in certain, mainly rural, areas of developing countries. This trend was reinforced in the 2000s, when renewables were further institutionalized in the agenda of the donor community. Particularly after the 2002 UN World Summit on Sustainable Development at Johannesburg, energy policies and renewables became prominent components for implementing national sustainable development strategies aimed at eradicating poverty and addressing climate change. Several programmes, combining financial and technical assistance, were launched on a bilateral level and multilaterally by UN agencies and international financial institutions.

Overall, until the late 2000s, green foreign energy policy focused on promoting the diffusion of renewables abroad, mainly within the context of (sustainable) development policy and climate objectives. Renewables were also included as an area of cooperation in bilateral and multilateral energy relations among consumers and/or producers, with an emphasis on research and policy exchanges. Examples include the US–EU Energy Council, the IEA, which expanded its agenda to cover renewables, and the IRENA. This latter organization, created in 2009, was promoted by a group of (mainly) European pioneering countries—especially Germany, which had tried without success (also because of opposition by the US) to establish renewable energy policy targets in the context of the UN—which were concerned that the IEA was institutionally captured by fossil fuel interests (Van de Graaf 2013; Urpelainen and Van de Graaf 2015).

In the aftermath of the 2008 global financial crisis and, more clearly, since the early 2010s, another green foreign energy policy driver has been added to the energy security–climate change–sustainable development package. In this period, renewables and low-carbon technologies began to be framed as one of the key frontiers for major powers' technological and industrial competition. Domestically, this prompted a new and more robust round of green industrial policy in North America, China, and Europe (Rodrik 2014). The expansion of green business in these regions and the increase in low-carbon technologies trade consolidated new actors' coalitions, calling for governments to act internationally to defend national business interests and/or promote favourable regulatory measures (e.g., Hughes, and Meckling 2017; Meckling 2018). International trade disputes over renewables, at the bilateral level or managed within the WTO, began in 2010 and escalated in the following years (Lewis 2014). Many of these involved China, which had become a leading green manufacturer, as complainant or respondent. The abrupt realization in Europe and the US that China was leading the field in several renewables industries and low-carbon supply chains—as a result of the same climate and energy policies they had implemented domestically—was a wake-up call for decision makers in Western foreign policy circles.

All these trends were strengthened in the second half of the 2010s. In 2015, the launch of the Agenda 2030 by the UN and the Paris Agreement additionally reinforced the donor community's focus on renewables for climate change mitigation, energy, and sustainable development objectives. The Covid-19 pandemic then dramatically exposed the vulnerability of Western powers with regard to shocks in several global supply chains, from medical equipment and drugs to microchips, semiconductors, and REEs. All the recovery measures implemented in Europe, China, and the US to tackle the economic downturn caused by the pandemic have incorporated a green dimension (e.g., Quitzow et al. 2021; Tian et al. 2021). This was intended to achieve climate and energy objectives and also economic imperatives related to industrial policy, growth, and major powers' competition over technological supremacy or sovereignty. In this context, critical low-carbon supply chains definitively emerged as a matter with national security implications. A development that was further accentuated after the Russian invasion of Ukraine, which triggered tensions in global energy markets amid a deterioration of West–China relations.

In sum, since the 1970s, various drivers have combined to induce national governments to develop a green foreign energy policy. Government intervention in this field, however, is more fragmented and less visible than is the case with traditional fossil fuel foreign energy policy. From an analytical perspective, there are four main issue areas that compose the building blocks of a green foreign energy policy: climate, energy, industry, and trade and investments. Each of these issue areas includes a subset of more specific operational objectives, which are often interrelated, that governments can pursue and that are connected to the wider strategic goals of green foreign energy policy. The climate area comprises those foreign policy actions that aim to promote mitigation measures in which renewables play an essential role. Supporting climate-mitigation policy and CO_2 targets—at the multilateral, regional, or bilateral level—entails indirectly supporting the diffusion of renewables, particularly for the decarbonization of the electricity sector. The energy issue area comprises several forms of international cooperation on renewables and all those measures that explicitly aim to promote renewables in third states. These latter measures can support various objectives, such as helping to meet energy demands in third states and promoting sustainable development. These measures, however, can also be implemented to achieve energy security and energy transition objectives for the senders because renewable energy produced abroad can be imported via cross-border electricity networks or hydrogen corridors.

The industry and trade and investment issue areas are strongly interconnected as well. The industry area comprises those measures aimed at supporting national green business internationalization and/or the creation of industrial partnerships and low-carbon supply chains in which national businesses have an important role. These measures illustrate the external dimension of the domestic industrial green policies, with which they are strongly linked. Indeed, in this area, the internal/external distinction is particularly blurred. The trade and investment issue area does not concern the trade of renewable energy per se, although it covers international trade in electricity and hydrogen generated by renewables. Moreover, this area concerns trade and investment in those critical raw materials, REEs, and technological equipment needed to produce and store renewable energy, such as solar PV panels, wind turbines, and lithium batteries (Sattich et al. 2021). Production technologies for renewables are often transferred internationally as intellectual property, for example, in the form of licensing, besides their tangible forms of manufacturing products. Hence, questions pertaining to intellectual property and foreign direct investments must be included in the trade and investment issue areas. Objectives typically pursued by governments in this area include the protection of domestic green industries; the expansion of trade quotas, market shares, and investment; and the protection of intellectual property rights over emerging renewable and low-carbon technologies. These objectives can partially overlap with climate-, energy-, and industry-related objectives. For instance, trade measures can be used to structure new low-carbon supply chains or protect domestic green industries from tariff and non-tariff barriers implemented in third states. Also, trade deals can be used as a vehicle for supporting the expansion of renewables abroad, such as in the case of trade deals that include provisions on climate or sustainable development. Trade and investment measures, however, are embedded in the existing international trade regimes, such as the WTO. These regimes are, at the same time, a constraint for the national green industrial policy—because they can limit the space to manoeuvre for governments—and an enabler or target of major powers' green foreign energy policy. They can act as an enabler because these regimes allow for international action to challenge domestic green measures implemented in third states. They can act as a target because promoting national regulatory preferences regarding renewable–energy–trade-related issues is another state objective in the context of the global energy transition.

Governments can pursue their green foreign energy policy objectives in the above-mentioned four issue areas by deploying a variety of tools. To organize these tools, it is useful to distinguish between three set of statecraft instruments: diplomatic, regulatory, and economic. Diplomacy involves traditional forms of interstate negotiations, communications, and cooperation but also transnational initiatives, public diplomacy, commercial diplomacy, and regulatory diplomacy. Commercial diplomacy can resemble Susan Strange's triangular diplomacy framework, which rests on classical, bilateral, government-to-government, and company-to-government interactions and views states as actively supporting their national companies in their pursuit of investments and contracts abroad (Stopford and Strange 1991). However, commercial diplomacy can also take the form of network diplomacy (Hocking 1999; Heine 2013). This notion highlights the increasing number of actors, ranging from local public and private actors to international financial institutions, that play a role in international economic relations and diplomatic exchanges. Like triangular diplomacy, network diplomacy focuses on 'doing things' rather than 'promoting rules'. However, the latter recognizes that states need the collaboration of a larger number of actors to achieve their objectives. Conversely, regulatory diplomacy sees governments or regulators negotiating and/or cooperating to promote rules and standards for economic transactions (for both goods and services) and market functioning. Regulatory diplomacy has grown in parallel with the rise of the regulatory state worldwide (e.g., Hocking 2004). The notion of the regulatory state underlines a particular conception of the state: the state as a regulator, i.e., as an entity that governs through regulations to achieve its objectives (e.g., Majone 1997; Levi-Faur, 2005). Governing through regulations means supplying and enforcing rules, typically in the form of command-and-control or market-based instruments. The regulatory state, however, has an external dimension (e.g., Vogel 1995; Bach and Newman 2007; Damro 2012; Lavenex, Krizic and Serrano 2017; Bradford 2012). The rules that it implements at home can have consequences in third states. Moreover, regulatory states actively promote international institutions, multilateral regulatory, and trade regimes or support the diffusion of their preferred rules in third states. This also includes 'regulating through trade' (Araujo 2019, 377), i.e., the use of trade deals that go beyond the removal of traditional barriers to trade and focus on promoting regulatory frameworks or regulatory harmonization. Indeed, in emerging technological fields, such as renewables and low-carbon technologies, states also compete to become rule-makers rather than rule-takers. Setting the standards for green industry, in particular, could 'lock in' favourable technological conditions and thus thwart competition (Wu and Chong, 2018, 508). Regulatory instruments differ from other traditional tools of economic statecraft (e.g., Baldwin 1985), which can be used as rewards (carrots) or punishments (sticks), such as foreign aid, investment guarantees, subsidies, preferential tariffs, and direct investment. In the context of the global energy transition, financial resources are very important to the realization of green investment projects abroad. These financial resources can be channelled through several economic tools, including state-owned companies, foreign aid packages or loans, and grants from national or multilateral development banks.

Diplomatic, economic, and regulatory instruments are common in many areas of international affairs. Although these instruments can be analytically separated, they are strongly intertwined in foreign policy. Another useful distinction emerges when we consider that these tools can be deployed along the entire spectrum, which illustrates the different modes of engagement in the area of foreign policy: unilateralism (without the approval of the counterparts), bilateralism, minilateralism, and multilateralism (Tago 2017). The unilateral–multilateral spectrum has analytical value, but it is worth stressing that foreign policies have always been a mixture of different '-isms'. A similar mixture is expected in the area of green foreign energy policy as well, although different international actors may place more emphasis on specific modes of engagement. Finally, geographical reach (regional or global) is another dimension of variation among green foreign energy policies. The same set of measures (e.g., foreign aid and trade deals) can be concentrated on a country's neighbourhood or deployed on truly global, intercontinental scale.

The four dimensions illustrated compose an analytical framework for mapping and studying green foreign energy policy. Conceptually, the measures in the area of green foreign energy policy can thus be collocated in the four-dimensional space illustrated in Figure 2.1.

Table 2.2 provides a compilation of examples to show how this four-dimensional framework can help empirically map green foreign energy policy measures. A similar framework is also useful because it allows us to highlight variations among international actors' actions on the matter and, thus, the comparative analysis of green foreign energy policies. Moreover, this framework allows for the empirical study of the evolution of green foreign energy policy over time. That is, it helps to trace policy changes in this field. These transformations can be collocated along a continuum indicating the magnitude of the shift: from a minor adjustment (i.e., in the intensity of the use of specific tools) to changes at the level of instruments (i.e., from economic to regulatory tools) and major changes, which imply a more profound shift involving both the tools and the overall approach of an international actor in terms of modes of engagement and the aim of its foreign policy (Hermann 1990; Hall 1993; see also Macikenaite 2020). Major changes in green foreign energy policy reflect transformations in green power strategies (see below).

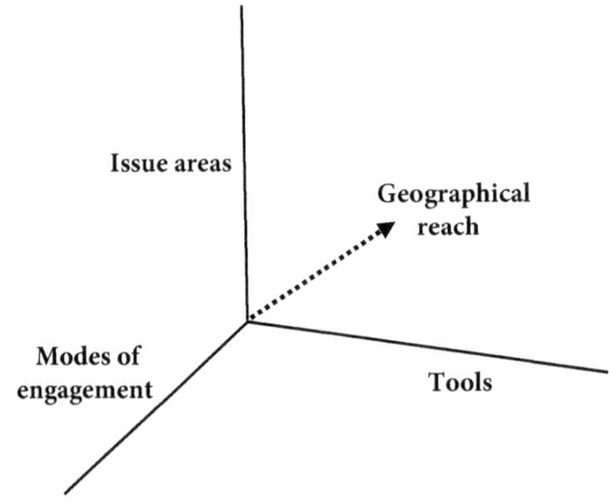

Figure 2.1 The four dimensions of green foreign energy policy

Note: The graphical representation of a four-dimensional space is usually made with a tesseract or tetracube, which is the four-dimensional analogue of a cube. Here, for clarity, only a sketch illustration of this figure is provided.
Source: Author's elaboration

Table 2.2 A matrix for mapping green foreign energy policy (with empirical examples)

Tools	Issue areas			
	Climate	*Energy*	*Industry*	*Trade and investment*
Diplomatic	China–ASEAN Environmental Cooperation Forum [Mi, R]	US–Canada–Mexico North American Energy Working Group [Mi, R]	US–China Renewable Energy Industries Forum [B, G]	EU–Namibia Partnership on Raw Materials and Renewable hydrogen [B,G]
	UNFCCC–Conference of the Parties (COP) climate change negotiations [M, G]	EU–Japan Green Alliance [B, G]	US–ASEAN Electric Vehicle Initiative [Mi, G]	US–Mineral Security Partnership [Mi, G]
	EU–China Partnership on Climate Change [B, G]	US–India Partnership to Advance Clean Energy [B, G]	US–Brazil Clean Energy Industry Dialogue [B, G]	US–DRC–Zambia Memorandum of Understanding for critical minerals supply chains [Mi, G]
Economic	UNFCCC–Green Climate Fund (national contributions to) [M, G]	EU–Neighbourhood Investment Facility supporting wind and solar projects in North Africa [Mi, R]	France's Engie green hydrogen partnership with the UAE's Masdar [B, R]	China's Molybdenum investments in Congo's copper and cobalt mines [B, G]
	China's South–South Climate Change Fund [Mi, G]	China's Silk and Road Fund supporting wind and solar projects in the Middle East [Mi, R]	Italy's Enel Green Power energy partnership with Morocco's Iresen [B, R]	US Development Financial Corporation investments in
	US Agency for International Development Climate Finance for Development Accelerator [Mi, G]	Just Energy Transition Partnerships with South Africa, Indonesia, Vietnam [Mi, G]	German–Australian Hydrogen Innovation and Technology Incubator [B, G]	Brazil's nickel and cobalt mines [B, G] Export–Import Bank of the United States (export credit for renewables' investments and trade) [B, G]

Continued

Table 2.2 *Continued*

Tools	Issue areas			
	Climate	*Energy*	*Industry*	*Trade and investment*
Regulatory	EU–Carbon Border Adjustment Mechanism [U, G]	EU–Energy Community Treaty [Mi, R]	G20 High Level voluntary principles on hydrogen (common standards for green hydrogen) [M, G]	EU–New Zealand Trade Agreement (with provisions on the respect of the Paris Agreement) [B, G]
	China–Green Investment Principles for the Belt and Road [U, G]	EU–European Economic Area (provisions on energy) [Mi, R]	US–International Partnership for Hydrogen and Fuel Cells (focus on regulations, codes, and standards) [Mi, G]	US–Japan Critical Minerals Agreement [B, G]
	Asian Infrastructure Investment Bank–Environmental and Social Framework (requiring projects to be in accordance with the Paris Agreement) [M, G]	ASEAN Power Grid Initiative [Mi, R]	EU–CertifHy (green hydrogen certification schemes) [U, G]	US–Mexico–Canada Agreement (provisions on renewables trade and investments) [Mi, R]

Notes: Modes of engagement: U = unilateral; B = bilateral; Mi = minilateral; M = multilateral. Geographical reach: G = global; R = regional. [...] = Modes of engagement, Geographical reach
Source: Author's elaboration

Internal–external nexus in green foreign energy policy

The previous section has clarified the building blocks of a green foreign energy policy, which are its main drivers and the dimensions that allow for the comparative mapping of this emerging field, as well as tracing changes over time. Exploring the internal–external nexus in the green energy transition is another key step in understanding variations and transformations in this field. Foreign policy analysis has traditionally investigated those internal political and economic factors that affect foreign policy choices (e.g., Gustavsson 1999; Hudson and Day 2007). Structural domestic changes—such as those triggered by the green energy transition—can alter the distribution of resources between powerful groups or constituencies, creating demands for specific courses of action in the area of foreign policy. Changes in political leadership, government agenda, public opinion, elites' perceptions, or bureaucratic preferences can also lead to transformations in the area of foreign policy. Recent work on economic statecraft, however, has pointed to the key role that state–business, or state–market, relations play in several areas of international affairs (e.g., Norris 2018; Gertz and Evers 2020; Aggarwal and Reddie 2020, 2021; Thurbon and Weiss 2021). This perspective has its roots in seminal IPE contributions from the 1970s and 1980s (for a review, see Underhill 2000). After all, much international economic activity (trade, investment, etc.) is actually conducted by companies, not by states. The domestic patterns of state–market relations cannot explain each single foreign policy decision-making process, but they illustrate the structural conditions within which foreign (economic) policy is conduit for a certain period of time. In particular, state–market relations are a constitutive component of state capacity (Weiss 1998; Cheung 2008). As scholars of historical institutionalism such as Krasner (1988) explain, these capabilities enable or constrain government action (see also Gertz and Evers 2020). State–market relations, however, are not fixed. They can change over time in response to both external and internal challenges even though they are embedded in specific institutional and ideational configurations pertaining to each country's national political economy (Clift 2014). These transformations, in turn, have implications for a state's approach to the management of international economic affairs.

In the realm of energy studies, the role of state–market relations and their external reach has always been very clear. For example, Randall (2005) described the US foreign oil policy from the Second World War to the 1990s, borrowing the IPE concept of forms of state and characterizing the US as an 'associational state' in this field. In an associational state, energy companies are in private hands and pursue their own short-term commercial interests. The government works with and for the companies and supports their business activities abroad. However, the government does not merely serve the oil industries but, rather, seeks to balance the companies' commercial interests with the country's long-term energy security interest (see also Goel 2004; Vivoda 2010). In the associational state model, private capital provides the financial resources and expertise, while the state provides political, diplomatic, and military support because it considers this to be an effective strategy via which to achieve its goals.

This pattern of state–market relations is closely linked to the history of the US energy sector and the country's broader approach to economic governance, which combines market liberalization and the defence of private rights with national security objectives. Those ideational and institutional factors prevented the development of something like a 'US National Oil Company' (Wälde 2008). A similar path was followed by several consumer countries, both in Western Europe, such as France, Italy, and the UK, and in Asia, such as Japan. The model of the 'partner state' can be used to describe this pattern of state–market relations in which governments create and nurture national champions at home and use diplomacy and other foreign policy tools to support those champions' activities abroad (Prontera 2017, 2019). When, with the liberalization and privatization of the energy markets initiated in the late 1980s, these patterns of state–market relations changed, those states' foreign energy policies were also affected (ibid.). In India and China, which are latecomers in the international oil and gas markets, foreign energy policy cannot be separated from domestic forms of state capitalism and the cooperative–competitive dynamics between government and national companies (Meckling, Kong, and Madan 2015).

State–market relations and their evolution over time are still important in understanding the various approaches to foreign energy policy in the low-carbon world. However, this internal–external nexus is simultaneously more significant and complex in the context of the global energy transition. It is more significant because acquiring a certain level of capacity in the area of renewables and low-carbon technology is a prerequisite for developing an active green foreign energy policy. It is more complex because this process entails government intervention in several interrelated sectors, such as industrial policy, energy policy, trade policy, and innovation policy. This stands in contrast to establishing national oil companies and granting them diplomatic support. Unlike the oil and gas industries, green business is characterized by a huge variety of companies that are active in various sectors and linked through global supply chains. This outcome reflects the intrinsic features of green energy business and the history of its development. Renewables began as a niche in fossil-fuel dominated energy systems. These niches have been nurtured, with important country-level variations in timing, means, and intensity, by government interventions, but for decades, they have been outside the radar of foreign policy. There was nothing particularly strategic or national security–related in green business until the early 2010s. Global supply chains in these sectors were not influenced by foreign policy imperatives but were, rather, driven by economic considerations.

The 2008 global financial crisis induced a worldwide rethinking of state–market relations. In the aftermath of the crisis, green industrial policies were significantly expanded through stimulus packages responding to the recession (Rodrik 2014). Several studies have explored the green turn taken by the developmental state in emerging economies and East Asia (e.g., Esarey et al. 2020). China, with its state-led (and party-led) model of capitalism and developmental approach, which is based on close state–business alliances at various levels of policymaking, has revealed an effective path towards renewables deployment and green industrial growth (Chen and Lees 2016). However, several works have demonstrated that the developmental state is not limited to emerging economies and East Asia but, rather, extends to Europe and the US, where the state has increased its role as an investor and developer of technologies and industries (Meckling 2018). This neo-developmental state differs from the classic developmental state. It operates through decentralized agencies embedded in networks, rather than a centralized bureaucracy, and through public–private partnerships and financial instruments to leverage private resources, rather than direct means of intervention, such as state-owned companies or banks (Ansell 2000; Ó Riain 2004; Block 2008; Weiss 2014; Mazzucato 2015). Such a neo-developmental network state also emerged in the US green energy sector (MacNeil 2013, 2014). Particularly, research has demonstrated how the US developmental state has expanded its activities to promote low-carbon technologies in order to address climate change, while regulatory measures failed (MacNeil and Paterson 2012; MacNeil 2013). The EU itself has begun to play a more active and interventionist role, mobilizing financial resources and promoting coordination and collaboration among state and non-state actors in order promote low-carbon technologies, energy networks, and green innovations (Prontera and Quitzow 2022). This move represented a shift with respect to the EU tradition as a regulatory state; although the bloc's developmental approach remains constrained by its nature as a quasi-state. Its evolution, hence, resembles what Lind (1992) indicated with the notion of a 'catalytic state'; i.e., a state that pursues its strategic objectives less by relying on its own resources than by acting through complex, multi-actor coalitions, transnational public–private consortia, and hybridized state–business alliances.

The diffusion of green industrial policies across industrialized and emerging economies after the 2008 financial crisis denotes a nascent green developmental state in the global energy transition. In this transformative process, the state is thus both a regulatory state, providing rules, and a developmental state, investing in industrial expansion and technological innovations. The structuration of this green developmental state at home—which takes different forms as it is embedded in different institutional and ideational configurations of each country's political economy—has not only contributed to shifts in state–market relations but also affected green foreign energy policy. Like the regulatory state, the developmental state has an external face. China has been a prime mover in this regard, attempting to replicate domestic state–green business alliances abroad (e.g., Liao 2021). However, a similar internal–external nexus is also evident in the case of the US and the EU. Under Obama, for example, Washington's green industrial policy built up an external reach. Whereas, in 2014–15, the EU established the Investment Plan for Europe, which focused on enhancing investment in climate-, energy-, and low-carbon-related sectors. Building on these innovations, soon afterward, in 2017, it launched the External Investment Plan, which added new financial instruments to the EU green foreign energy policy arsenal.

The trend towards the global expansion of the green developmental state has been further reinforced in the wake of the Covid-19 pandemic. Again, the recovery measures implemented in Europe, the US, and China—as well as in several other countries—included a green component targeting renewables and low-carbon technologies. Moreover, they took the form of interventions expected by a developmental state. The bulk of the green measures included in the post-Covid-19 recovery packages implemented by the larger emerging and industrialized economies represented grants or loans (about 40% of the total) and tax reductions or other subsidies (19% of the total), while regulatory innovations were only around 14% of the total (OECD 2021). These green recovery measures, however, were enacted in a very different context. As illustrated, the Covid-19 pandemic had revealed problems related to supply chains vulnerabilities in critical sectors. This revelation intertwined with the growing geopolitical and geoeconomic competition among major powers, which has been aggravated by the Russian invasion of Ukraine. In a similar context, state green interventionism acquired a new meaning. Conventionally, the concept of statecraft is applied externally via foreign policy measures. However, economic statecraft can be 'domestically oriented' when state intervention in industrial and technological sectors is not driven only by economic objectives—as in the traditional developmental/neo-developmental state perspective—but, rather, aims to respond to challenges arising from the international environment, whether such pressures are geopolitical or geoeconomic in nature (Weiss and Thurbon 2021; Weiss 2021). This form of economic statecraft in the area of renewables and low-carbon technologies has become evident in the aftermath of the Covid-19 pandemic and the Russia–Ukraine war, triggering a further blurring of internal/external measures. Overall, the recent developments call attention to how the ongoing restructure of state–market relations—in a developmental, neo-developmental, or catalytic state fashion—affects the major powers' wider approach to green foreign energy policy, i.e., their green power strategies.

Green power strategies

In Chapter 1, I discussed the features of green superpowers in terms of their control over key positive and negative power's resources for the global energy transition. Unlike this more static perspective, the notion of power strategies highlights a process-oriented understanding of power in international relations (e.g., Keohane and Nye 1977), i.e., the way in which states attempt to use the resources and capabilities at their disposal to achieve their goals when interacting with other states. Green power strategies illustrate the efforts of major international actors to achieve the strategic goals of green foreign energy policy while increasing their posture and influence in the emerging low-carbon world. Indeed, as highlighted, the global energy transition implies a reconfiguration of power relations and the development of new alliances, partnerships, and green trade links. Green power strategies are affected by the resources and capabilities—including in terms of state–market relations—preferences, and ambitions of international actors and by the opportunities offered and the constraints imposed by the internal and external environments. These pertain both to the constellations of power and interests involved and the institutional landscape at the domestic and international level. In the fossil fuel world as well, major powers' energy security strategies were affected by such factors. As illustrated, because of its military capabilities, the US could develop an energy security strategy with an important military edge. Washington could also engage globally with key producers, establishing strategic partnerships, as in the case of

US–Saudi relations. Moreover, because of the size of its domestic oil market, the US could influence international energy affairs through unilateral measures (e.g., Ikenberry 1988). Other large consumers in Western Europe or in Asia could not rely on the same capabilities to promote their energy interests abroad. Instead, they mobilized their economic power—often in the form of state-owned companies—and resorted to bilateral relations and commercial diplomacy to achieve energy security objectives.

In general, major powers are those that can more easily resort to bilateral or unilateral modes of foreign policy. However, the differences along the unilateral–multilateral spectrum also incorporate a value-based perspective. Each mode of engagement is supported by some principles prescribing what kind of international order is most desirable (Ruggie 1992; Tago 2017): one based on diffuse reciprocity and equality among states (multilateralism), one based on smaller clubs of like-minded allies (minilateralism), one based on case-by-case preferentialism (bilateralism), and one legitimizing a self-centred foreign policy for powerful states (unilateralism). The choices regarding the favoured mode of engagement thus reflect not only different resources and capabilities but also different principles and approaches to international affairs. For instance, the Energy Charter Treaty, launched by the EU in 1994, was imbued by the bloc's preferences for a multilateral and rules-based international order. The level of development of the institutional landscape on a particular issue can also favour the conduit of foreign policy within multilateral settings. If dedicated institutions are already in place, it can be easier—and less costly—to promote national goals through these institutions rather than through ad hoc unilateral actions or bilateral deals. Similarly, international institutions can act as constraints on national choices, increasing the costs and delegitimizing unilateral or bilateral strategies. Differences in capabilities can also explain variations of green power strategies in terms of their geographical reach. Only a few international actors have those assets needed to develop a foreign policy with a global reach. However, a regional dimension can also be preferred for other reasons in the area of green foreign energy policy. Geographical proximity and sectoral features can support similar choices. For instance, consumers can focus their external action on targeting neighbourhoods that have export potentials in the area of renewable energy through regional cross-border electricity networks or hydrogen corridors.

Four conceptualizations of power—linked to long-standing perspectives in international relations theories—can be used, however, to better characterize the green power strategies of major international actors: soft power, regulatory power, economic power, and catalytic (or collaborative) power.

As with foreign policy tools, states normally combine these assets in their power strategies, although with a different emphasis. Hence, these conceptualizations can help us to understand the approaches to the global energy transition adopted by major international actors, from both a comparative perspective and diachronically, to make sense of their evolution over time. Soft power is linked to intangible assets, such as culture, political values, institutions, and policies (Nye 2004, 2008, 2011). These resources can influence others when they perceive the authority and legitimacy of an actor and trust its ideas (Adler 1997; Finnemore and Sikkink 1998). Success at the domestic level can provide a country with additional 'moral agency' (Davies and True 2017) in promoting its policy solutions to critical issues and 'leading by example' (Oberthür and Roche Kelly 2008). In some cases, this can mean that actors may obtain what they want, even without deliberate action to promote particular ideas or policy approaches. Soft power, however, can be deployed deliberately to influence others. The means employed in this regard represent the second dimension that distinguishes soft power from the harder version of power. Soft power does not involve rewards or punishment; it is wielded 'through the cooptive means of framing the agenda, persuading, and eliciting positive attraction' (Nye 2011, 13). More specifically, this manifests through diplomacy, institutional venues for communication and dialogue, and capacity-building efforts to foster policy learning and diffusion, which are all important instruments for mobilizing soft power resources (Nye 2008; Quitzow and Thielges 2022). Soft power, however, mainly works indirectly. It aims to influence others by shaping, with a long-term perspective, the milieu in which their behaviour take place (e.g., Nye 2004; Kearn 2011; Nielsen and Vilson 2014). This is the third difference between soft power and other power strategies that attempt to alter behaviour in the short term. The notion of soft power was initially developed with a Western bias and its liberal international agenda. However, it has since been applied to other international actors, including China (e.g., Mingjiang 2008; Li et al. 2016).

Regulatory power relies on tangible resources and different tools and channels of influence than soft power (e.g., Vogel 1995; Bach and Newman 2007; Damro 2012; Bradford 2012, 2020; Lavenex 2014; Goldthau and Sitter 2015a; Lavenex, Krizic, and Serrano 2017; Lavenex, Serrano, and Büthe 2021). This conception of power is linked to the literature on the regulatory state and its global diffusion (e.g., Lavenex, Serrano, and Büthe 2021; on the rise of regulatory state in China, see Bach, Newman, and Weber 2006; Hopewell 2021). Key assets for deploying regulatory power are market size and regulatory capacity, i.e., the capacity to formulate, implement, and uphold international rules (e.g., Bach and Newman 2007; Bradford 2012, 2020; Damro 2012;

Lavenex, Krizic and Serrano 2017). Overall, regulatory power stresses how influence in international economic affairs is related to the attractiveness and size of a state domestic market and its ability deploy its regulatory toolbox; promote rules, institutions, and regulatory regimes; and engage in regulatory diplomacy. In the energy realm, regulatory power is close to an international liberal agenda (e.g., Goldthau and Sitter 2015a, 2015b). This is particularly the case when regulatory power is wielded in multilateral settings and through international institutions. However, regulatory power can also have a hard hedge, placing this notion closer to the realist-mercantilist conception of economic power. This occurs when states adopt regulations 'selectively', using an approach that targets specific market actors (e.g., Goldthau and Sitter 2015a, 2018). That is to say, a state regulatory toolbox, in this case, is not deployed to pursue liberalization, setting the 'rules of the game', or a level playing field for market actors but, rather, in order to achieve particular geopolitical goals. A case in point in the fossil fuel energy world has been the deliberate application of the EU's liberal energy market rules to challenge the Nord Stream pipelines project promoted by the Russian Gazprom (e.g., Goldthau and Sitter 2018; Siddi and Kustova 2021).

As suggested, this use of states' regulatory tools is closer to a mercantilist-realist understanding of international relations. This perspective is particularly embraced by the economic power conception (e.g., Gilpin 2011). Economic power—which relies on the combination of several national economic resources—implies the mobilization of state economic might through more direct means, such as state-owned companies and banks. Moreover, this power strategy is accompanied by traditional forms of commercial diplomacy that view governments as backing national businesses and promoting investments in third states. Economic power, however, also includes the more traditional use of economic tools for reward (e.g., subsidies, investment guarantees) or punishment (e.g., trade tariffs, embargoes). These are not aimed at promoting international rules and standards but, more directly, at promoting national interests. In this regard, economic power typically unfolds in bilateral, rather than multilateral, modes of engagement. Finally, economic power is associated with the ability of major international actors to establish spheres of influence with friendly states (e.g., Etzioni 2015; Allison 2020). In this sense, a strategy based on economic power can be supported by military power because such military power can be deployed, along with economic means, to create strategic alliances and design spheres of geopolitical influence.

Like economic power, catalytic power is not concerned with promoting rules but, rather, mobilizing financial resources and supporting investment projects abroad. However, catalytic power relies on economic rewards rather than punishment, it is upheld by network diplomacy rather than traditional commercial diplomacy, and it is closer to a liberal approach to international economic affairs (Prontera 2019; Prontera and Plenta 2020; Prontera and Quitzow 2023). This conceptualization of power is linked to Lind's model of the catalytic state (Lind 1992; see also Prontera 2020; Prontera and Quitzow 2022). A key notion of this model is that states seek to extend their capabilities though collaboration with other actors and hybrid public–private alliances. In this sense, catalytic power resonates with the concept of 'collaborative power' (Slaughter 2011, 2017; see also Weiss 2014). This form of power relies less on actors' own resources and more on mechanisms of mobilization and connectivity for creating multi-actor coalitions to promote state objectives. In this vein, this conceptualization of power extends beyond a narrow linkage with the catalytic state model. As suggested by this model, however, financial instruments can serve as a catalyst to mobilize action by other actors (Prontera and Quitzow 2023). Connectivity, on the other hand, relates to the broader notion of network diplomacy, which Hocking (1999) also explicitly indicates as 'catalytic diplomacy'.

Building on this discussion—and considering the strategic goals of green foreign energy policy—it is thus possible to characterize four green power strategies (Table 2.3).

Table 2.3 Conceptualizations of power and green power strategies

	Green Power Strategies			
	Soft power	**Regulatory power**	**Catalytic power**	**Economic power**
Aim	Shaping the milieu	Promoting international rules and standards	Creating multi-actor coalitions	Promoting national interests
Means	Ideas and policies, leading by example, diplomacy, institutional venues for communication and policy exchange, capacity building and technical assistance	Regulations, international institutions and regulatory regimes, regulations through trade, regulatory diplomacy	Economic rewards, financial tools for mobilization, hybrid alliances and public–private partnerships, network diplomacy	Economic rewards/ punishment, state-owned companies and banks, regulations applied 'selectively', commercial diplomacy

Source: Author's elaboration

A green soft power strategy aims at shaping the milieu in which the global energy transition takes place. It relies on intangible assets and means in terms of ideas and policies, as well as on diplomacy, limited financial support for capacity building or technical assistance, and institutional venues for communication, dialogue, and policy exchange (see also Quitzow and Thielges 2022). This strategy is less concerned with promoting national green business and does not focus on reducing vulnerabilities along low-carbon technology supply chains. However, it can provide states with channels of influence over the forms and content of the energy transition and, thus, improve their posture in the emerging low-carbon world. A green regulatory power strategy relies on tangible assets, such as market size, and those capacities that constitute the institutional core of the regulatory state. This strategy aims at designing the 'rules of the game' for the global energy transition, which can, at the same time, support national green business, orient green trade patterns, and aid in the diffusion of renewables and low-carbon technologies. When this strategy is deployed in multilateral modes, it supports a liberal green international agenda. Conversely, economic green power illustrates a strategic approach with a mercantilist edge. This strategy is less concerned with shaping the norms and rules for the energy transition. Rather, it focuses on promoting national green business, reducing vulnerabilities along low-carbon technology supply chains, and using economic might in a more direct way to increase a state's influence in the low-carbon world. This green power strategy tends to manifest through bilateral modes of engagement, although it can be deployed to create green spheres of influence. Moreover, this strategy can be associated with military power and affect whether new alliances and strategic partnerships are formed (or strengthened) with a green energy component attached to them. A green catalytic power strategy can also be used in promoting national green business, reducing vulnerabilities along low-carbon technology supply chains, and supporting the spread of renewables. However, unlike green economic power, this strategy relies on economic rewards and collaborative power. This strategy is closer to a liberal approach to green energy relations, although it relies on different means than regulatory power. A green catalytic power strategy aims at creating coalitions to support green investments and projects rather than at promoting the 'rules of the game' for the energy transition. This strategy tends to unfold in minilateral modes of engagement because creating coalitions of larger actors is challenging. State–market relations are also important for this green power strategy. However, they tend to take the form suggested by the catalytic state model: hybrid alliances and public–private partnerships, rather than state-owned companies or more direct means of mobilizing states' economic might.

3
China
The Unexpected and Contested Green Giant

From energy security to technological dominance

After the Second World War, China was for a long period essentially self-sufficient from an energy point of view. Coal, in particular, was the main resource exploited by the leaders of the Chinese Communist Party (CCP) to drive the country's economic development and to ensure the energy supply for its vast population. China's large coal reserves fuelled the rapid industrialization and urbanization process initiated since the beginning of the Deng Xiaoping era, in 1978, and its policy of reform and opening up to the world. Today, coal still represents the backbone of the Chinese energy system, especially in the electricity sector, where it accounts for 61% of production, well ahead of hydropower (15%), wind and solar (approximately 14%), nuclear energy (4,7%), and other sources (Table 3.1). As a result of this inward-looking energy strategy, China has been on the margins of the main events that have characterized global energy politics in the twentieth century, such as the oil shocks of the 1970s and the diplomatic and military involvement of the great powers in the Middle East. This began to change after the end of the Cold War, in parallel with China's dramatic economic development, which meant that domestic resources could by no means meet the country's fast-growing energy demand. In 1993, China joined the club of oil-importing countries. Oil consumption continued to grow, and in 2014 China became the world's largest oil-importing country, assuming the position that had long been occupied by the US. In the natural gas sector, consumption surpassed domestic production in the early 2000s and has rapidly increased since then.

This new reality of energy dependency on external supplies was soon perceived as a potential weakness for a country accustomed to self-sufficiency and that was a latecomer in international energy markets. Energy security has since become a major concern for policymakers in Beijing. Their response followed a path not unlike that originally taken by consumer countries in Western Europe. The Chinese government promoted a reorganization of

Green Superpowers. Andrea Prontera, Oxford University Press. © Andrea Prontera (2024).
DOI: 10.1093/9780191987304.003.0003

Table 3.1 Share of electricity production (percentage) by source in China (2000–2022)

	2000	2005	2010	2015	2020	2022
Coal	78.2	79.2	77.2	70	63.6	61.3
Hydropower	16.4	15.8	17	19.2	17	15
Oil	3.4	2	0.35	0.17	0.14	0.14
Gas	0.4	0.4	1.8	2.8	3.2	3.1
Nuclear	1.2	2.1	1.7	3	4.7	4.7
Wind	0.06	0.08	1.18	3.2	6	9.1
Solar	0.01	0.01	0.02	0.6	3.3	4.7
Bioenergy	0.1	0.02	0.5	0.9	1.7	1.9

Source: Author's elaboration from Ritchie, Roser, and Rosado (2023)

the national energy sector that resulted in creating three giant, vertically integrated, state-owned oil companies: CNPC, Sinopec, and CNOOC. Beijing went on to support the internationalization of these companies with active, bilateral diplomacy backed by the state, which coordinated its energy objectives with other goals of its 'going-out' policy, formally initiated in 1999. With this approach, the Chinese government aimed to secure supply contracts and energy imports from important producers in the Middle East (such as Saudi Arabia and Iraq), Africa, Latin America, and Central Asia. Several studies have analysed the dynamics, potentials, and consequences of this Chinese fossil fuel foreign energy policy based on a direct mobilization of the growing state economic power and diplomatic clout (e.g., Kong 2009; Currier and Dorraj 2011; Andrews-Speed and Dannreuther 2013; Tunsjø 2013; Zhao 2014).

While concerns over China's energy security have not disappeared—and the country consolidates its position in the geopolitics of fossil fuels in the twenty-first century—the CCP's leadership has expanded and reframed its approach to this matter. Renewables and low-carbon technologies, in particular, have begun to play a decisive role in addressing multiple challenges. These include energy security, as China can rely on abundant renewable energy resources, especially in the north and south-west of the country, which in the long run can significantly reduce its reliance on imports. These challenges also include environmental protection, climate change, economic development, and technological dominance. This shift was initiated in the late 1990s and escalated in the following decades. Hydropower has always been important for China; for several years, the construction of large hydroelectric dams has been a priority for Chinese planners, and this source maintains an important position in the country's energy mix (Table 3.1).

By contrast, apart from very small projects in rural areas, often supported by Western-led multilateral development banks and bilateral aid agencies, the diffusion of wind and solar energy begun only in the late 1990s. The initial motivations for promoting these technologies were rural electrification and poverty alleviation (Pan et al. 2006). However, a growing awareness of the negative health and environmental consequences of coal usage also become an important driver for renewables at that time. In 1996–1997, the government launched the Riding the Wind Programme and the Brightness Programme for wind and solar installations. However, these programmes had only a limited impact, due to inadequate coordination between the various agencies involved in their implementation (Andrew-Speed and Zhang 2019). Moreover, the leadership of the CCP remained strongly focused on promoting economic development and growth, even at the cost of environmental damage. This approach also informed China's position on global climate change policy.

China participated in the 1992 UN Rio Conference and in the negotiations for the UNFCCC. This diplomatic engagement allowed the country to alleviate the tensions that emerged with the West in the wake of the Tiananmen events, signal its respect for international environmental norms, and promote its further integration into the international community after the end of the Cold War (Chen 2008). However, China's multilateral climate diplomacy was guided by a conservative position aimed at safeguarding the country's right to develop under the 'common but differentiated responsibilities' (CBDR) principle (e.g., Chen 2008, 2012; Qi and Wu 2013; Stensdal 2014; Yang 2022). From the outset, China's leaders framed climate change as a Global North vs. Global South issue, representing the country as part of the developing world that should prioritize economic growth and poverty eradication over environmental protection. China's leaders also stressed that international cooperation on climate change should be based on the respect for national sovereignty, that is, with limited intrusiveness on domestic choices or binding commitments. China maintained this position during the 1990s and promoted the CBDR principle by leading a large coalition of developing countries under the G77–China grouping (e.g., Chen 2012; Sandalow et al. 2022). In 1997, Beijing joined the Kyoto Protocol, which imposed emission limits on industrialized countries, but not on developing countries such as China. In this period, China also increased bilateral diplomatic relations with several states on renewable energy technology and climate issues (Lewis 2023a). However, the country was still a 'taker' of policies and innovations developed by other (mostly Western) pioneering countries. It was only later that China was able to shape the global energy transition with its foreign engagements. This was to be the result of incredible and unexpected internal developments in the area of renewables. In the late 1990s, experts from the US Department of Energy who were helping China to exploit its renewables potentials correctly affirmed that this sector in the country was still 'in an early stage' (Wallace, Jingming, and Shangbin 1998, 131). A decade later, China would be a manufacturing powerhouse for solar and wind technologies, on the way to surpassing Europe and the US in terms of renewable capacity installed.

In the early 2000s, several developments combined to accelerate the diffusion of renewables in China (Chen 2012; Lewis 2017; Andrew-Speed and Zhang 2019; Gang 2019; Nahm 2021). In 2002, Beijing ratified the Kyoto Protocol and it actively began to take part in the clean development mechanisms (CDM), especially with a view to promoting renewable energy projects. Air pollution was a huge problem in several Chinese cities, and addressing this issue was now important for the legitimacy of the CCP. Chinese planners focused effectively on renewables as both a contribution to tackle environmental problems and as an industry with important growth potential. Meanwhile, a round of government reforms between 1998 and 2003 resulted in a progressive centralization of energy policymaking in the hands of the National Development and Reform Commission (NDRC)—previously the State Development Planning Commission—which works under the State Council as the main agency for managing China's economy. In 2003, the NDRC launched a national programme for wind installations—the Wind Power Concession Programme—which included a local content rule to sustain Chinese wind manufacturers (this rule was abandoned only in 2009 as it was deemed to be in conflict with WTO provisions). This move inaugurated a gradual convergence of the country's renewable energy policy and innovation and industrial policy, a trend that was to become more evident under the 'indigenous innovation' strategy put forward during the 11th Five-Year Plan (FYP; 2006–2011) and even more so under the 12th FYP (2011–2015). This trend was to result in the structuring of the Chinese green energy state—that is, an ensemble of policies and institutions that actively supported the diffusion of renewables and the country's low-carbon technology and industrial system. Unlike in Western countries, where the growth of renewables occurred in the context of liberalized and privatized energy markets, China's trajectory followed from the outset a state-led and developmental approach—an approach that was, however, embedded in the Chinese party-led model of capitalism and 'fragmented' authoritarianism (Lieberthal and Oksenberg 1988; Grünberg 2017). This approach, which was upgraded after the 2008 global financial crisis, proved to be very effective. In the following decade, China surpassed the EU and the US in wind and solar installed capacity (Figure 3.1, Figure 3.2). Moreover, Chinese wind turbine (e.g., Goldwind, Envision, Ming Yang) and PV (e.g., Jinko Solar, JA Solar, Trina Solar, Risen Energy) manufacturers climbed the global rankings, as the country became the main world's producer of these technologies.

During the 2000s, the Chinese government was also concerned about ensuring critical minerals in order to sustain the growth of the country's green manufacturing capacity. This concern was important even in the case of electricity grids that required consistent expansion to meet a fast-growing demand. China is endowed with a number of important mineral resources, especially REEs. Since the 1990s, policymakers in Beijing had realized how central they were for national economic development and extended state control over the sector (Kalantzakos 2017a). The government not only supported mining activities, but it also promoted the processing industry; a business with high environmental impact that several Western countries were happy to outsource. However, to keep pace with burgeoning internal demand, imports were required as well. Consequently, under its 'going out' policy, the Chinese government increased its support not only to favour the internationalization of its fossil fuel companies, but also that of other state-owned enterprises (SOEs) and private companies involved in the minerals and mining sector. With the strong backing of the Chinese apparatus for

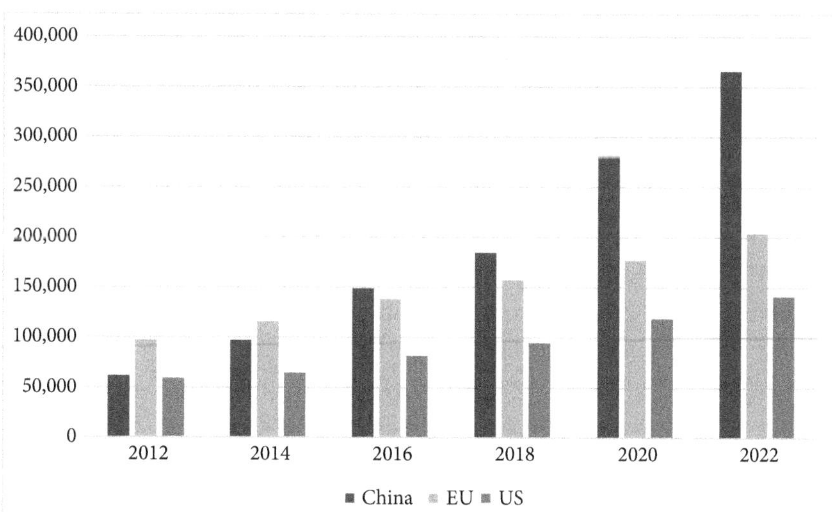

Figure 3.1 Wind energy installed capacity in China, the EU, and the US (in MW, selected years)

Source: Author's elaboration from IRENA (2018, 2023a)

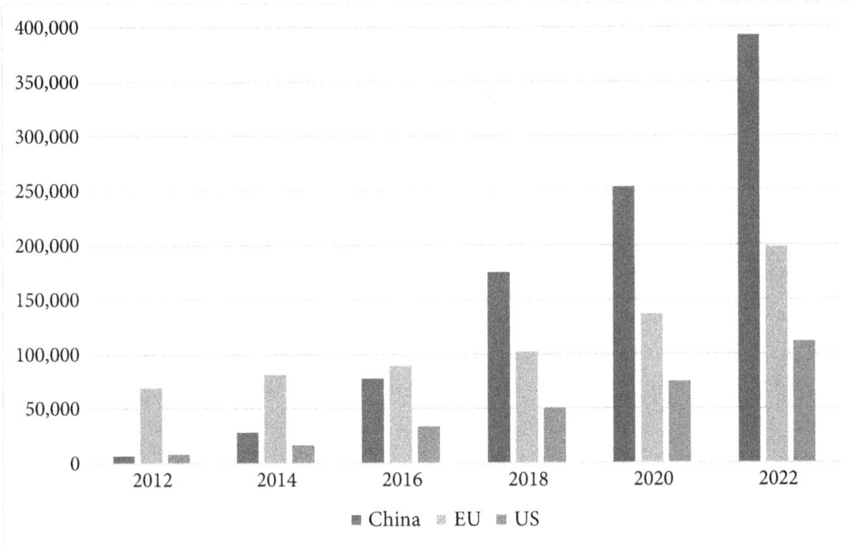

Figure 3.2 Solar PV installed capacity in China, the EU, and the US (in MW, selected years)
Source: Author's elaboration from IRENA (2018, 2023a)

economic statecraft, such as the Ministry of Commerce (MOFCOM), which until 2018 was also responsible for the country's foreign aid, as well as the state-owned policy banks—that is, the China Development Bank (CDB) and the Export–Import Bank of China (CHEXIM), which are tasked with supporting the government's domestic and overseas development agenda—these companies invested abroad to secure access to several mineral resources, including cooper, cobalt, lithium, and nickel. In South East Asia, for example, Chinese companies invested in mining projects in Indonesia (nickel), Myanmar (nickel and copper), Papua New Guinea (nickel and cobalt), the Philippines (nickel), and Vietnam (copper; Andrews-Speed, Qiu and Len 2016). Moreover, Chinese mining activities developed on a truly global scale, specifically targeting regions of the Global South such as Latin America and Africa. In the Democratic Republic of Congo (DRC) Chinese companies, such as the mining giant Molybdenum, arrived to dominate cobalt production (e.g., Gulley, McCullough and Shedd 2019). Although it might seem paradoxical, the first sign of the Chinese green foreign energy policy can be identified in this 'going out' in the critical minerals sector. Obviously, the 'going out' policy was anything but green; it was built primarily around fossil fuels, and Chinese mining activities abroad had severe negative environmental consequences (e.g., Shinn 2016). However, the centrality that China has acquired in the supply chains of the critical minerals required for the current global energy transition is as a result of activities undertaken in that original period. These actions focused on industry, trade, and investments on a global scale, and relied on the deployment of economic power and bilateral commercial diplomacy.

The 12th FYP (2011–2015) also signified a key change for China's energy policy priorities (Meidan 2020). As proclaimed by Prime Minister Wen Jiabao, this plan aimed at addressing the four 'un's' and reorient an economy that was seen as 'unstable, unbalanced, uncoordinated, and unsustainable' towards a 'new normal' trajectory of development. This reorientation was intended to shift the Chinese economy towards innovation and services rather than heavy industry, but it was also in line with ongoing domestic concern about environmental degradation and air pollution. In 2009, ahead of the COP 15 in Copenhagen, China announced its first ever goal concerning CO_2, pledging to lower carbon intensity—the ratio of CO_2 emission per unit of GDP—by 40%–45% from 2005 levels by 2020. However, China did not reverse its approach to global climate negotiations (Wu 2016; Yang 2022). It insisted on its identity as a developing country, aligning itself with other emerging economies—the BASIC coalition with Brazil, South Africa, and India—to defend the CBDR principle and reject binding commitments on its emissions. This Chinese approach contributed to the quasi-failure of the Copenhagen conference. However, it was in this period that a second component of the Chinese green foreign energy policy emerged, as the government increased its support for the internationalization of Chinese green business. This move was driven by different motivations: helping Chinese companies to acquiring new technologies and increase exports in the wake of overcapacity in the domestic market, as well as supporting the realization of wind and solar projects abroad in response to demands for similar investments from developing countries. In any case, this new building block of the Chinese green foreign energy policy followed the previous approach based on the direct mobilization of Chinese economic power and (bilateral) commercial diplomacy. By contrast, Beijing did not deploy regulatory tools to foster the energy transition. China's approach focused on the externalization of its green developmental state model rather than on promoting regulatory regimes or policy changes abroad. This is well exemplified by China's adherence to the 'host country standards' in its foreign energy, environmental, and economic engagements, as well as in its unwillingness to use trade deals to promote rules in third countries (e.g., Voituriez, Yao, and Larsen 2019; Shaffer and Gao 2020).

National energy policy priorities continued to evolve in the 2010s. In 2012, the notions of 'green economy' and 'ecological civilization', which was linked to the country's traditional ecological philosophy, were officially embraced by the Chinese leaders at the 18th Congress of the CCP (Qi and Wu 2013). These ideas and China's green energy domestic and foreign policies were further advanced with the coming to power of President Xi Jinping. China, which was not part of the IEA—the energy club of the OECD countries—joined the IRENA in 2013, signalling Beijing's growing interest in renewable energy. China also became part of high-level multilateral global forums, promoted by the US, for green technology cooperation and development, such as the Clean Energy Ministerial and Mission Innovation (see Chapter 5). More importantly, in 2013, Xi Jinping launched the Belt and Road Initiative (BRI), which built on the 'going out' policy to escalate the economic, industrial, and diplomatic engagement of China in several regions of the world. Then, in 2015, the Chinese government issued the Made in China 2025 initiative. This initiative, together with the 13th FYP (2016–2020), confirmed the new focus of the CCP leadership on an innovation-driven development strategy in which renewable energy and emerging low-carbon technologies (e.g., EVs, batteries, and smart power grids) should play a key role. Since this period, China's investment in the energy transition has dwarfed that of the other major powers (Frankfurt School-UNEP Centre/BNEF 2020). This occurred while control of the CCP over the state, as well as of the party–state over the market and the national economy, was strengthened under Xi Jinping; for example, with many SOEs and large private enterprises being obliged to increase links with Party committees (Grünberg and Drinhausen 2019).

In 2016, the Chinese Ministry of Natural Resources issued the National Plan for Mineral Resources (2016–2020), which confirmed that ensuring supply of minerals for emerging industries was a government priority that required strict state control. The plan created a catalogue of strategic minerals—including copper, nickel, cobalt, lithium, and REEs—and encouraged international cooperation, especially in the context of the BRI, China–Latin America, and China–Africa relations. In the area of renewable energy policy, the 13th FYP (2016–2020) signified a period of gradual phasing out of subsidies for the deployment of wind and solar installations—now cost competitive with fossil fuels—and a shift towards the use of more market-based instruments. This shift will continue under the 14th FYP (2021–2025) and has included the 2021 launch of the Chinese national Emission Trading System (ETS), which built on previous experience at the local levels. However, these innovations did not open the way for major market reforms in the power or carbon market (Meidan et al. 2023). The CCP and the state have continued to retain their grip on energy policy, including through state ownership, and to directly steer green industrial and innovation policy. The goal of the government was now to maintain technological supremacy over well-established sectors, such as PV and wind energy, and to expand this dominance over new sectors such as EVs, batteries, smart grids and, later, green hydrogen.

Under Xi Jinping, China finally also changed its approach to multilateral climate negotiations, deciding to play a more constructive role (Hilton and Kerr 2017). In anticipation of the COP 21 held in Paris, China pledged to peak CO_2 emissions by 2030 and engaged in bilateral diplomacy with key players, such as the EU countries, India, and, most notably, the US (Yeophantong and Goh 2022). China also engaged several countries of the Global South both bilaterally and through the BASIC and G77–China groups. This new position adopted by China was instrumental in ensuring the success of the COP 21. Importantly, China's position did not change even when the US, under the Trump administration, decided to withdraw from the Paris Agreement. On the contrary, China continued to sustain the post-Paris global climate regime and to improve its efforts in the area of mitigation. However, China did not seek to assume the mantle of a global climate leader, as the CCP did not want to risk its economic goals or be constrained by binding obligations (Qi and Dauvergne 2022). Beijing was to maintain this position in subsequent global multilateral meetings, from those under the UNFCCC to the efforts in forums such as the G20. This position is in line with broader Chinese foreign policy since the 'reform and opening-up' era, which is guided by principles of non-interference and sovereignty, with multilateral interactions serving primarily domestic policy goals (Qin 2014; Sutter 2016).

Despite this continuity, the post-Paris period saw the beginning of a more consistent process of greening China's external energy relations. This manifested in Chinese bilateral and regional engagements, in the upgrading of its climate finance and, most notably, in efforts to green the BRI, which under Xi Jinping became China's main foreign policy framework for engaging the Global South (it was inserted in the CCP constitution in 2017), and which, like the 'going-out' policy, had been based largely on fossil fuels. These innovations expanded the scope of the Chinese green foreign energy policy, from critical minerals and supporting national green business to climate finance and other means of influencing and shaping the global energy transition; the country had turned from a taker to a maker able to offer technologies and policy ideas, especially to the Global South. With its efforts aimed at greening the BRI, China has also sought to enrich its green power strategy, adding soft power and regulatory power dimensions to its traditional economic power perspective.

After the Covid-19 crisis, China continued to strength its position on renewables. Its post-Covid-19 recovery package provided approximately $20 billion spending in support of the green energy sector (IISD 2021). However, as a result of the pandemic crisis, which resulted in a global economic downturn and a more challenging international environment, the CCP leadership formulated the so-called 'double circulation' strategy (Herrero 2021). This strategy, which became part of the 14th FYP (2021–2025), focused on recalibrating China's industrial policy, placing a renewed emphasis on state-led growth and self-sufficiency based on the country's vast domestic market and its manufacturing and innovation capacity, including in the area of low-carbon technology. At the same time, with the China Standard 2035 initiative—an ideal continuation of the Made in China 2025—the government increased its efforts to become a global standard setter in emerging technology, engaging the system of international standards development organizations. This represented a shift in relation to the previous period, in which the diffusion of Chinese technological standards was pursued primarily through the overseas investments of Chinese companies or through the network of commercial and trade agreements of the BRI framework (Seaman 2020; Shaffer and Gao 2020).

In the field of climate policy, the government committed itself to two important goals: achieving peak CO_2 emission before 2030 (an upgrade of the pre-Paris commitment) and carbon neutrality by 2060. The carbon neutrality pledge, announced in September 2020 in a speech by President Xi Jinping at the UN General Assembly, implies a dramatic transformation of China's energy system still overwhelming dependent on coal. This resulted in the 2021 launch of the so-called '1+N' policy framework, which provides for a wide set of initiatives aimed at achieving the new climate goals (e.g., circular economy, low-carbon transport, energy savings, urban and rural development), including an acceleration in the deployment of renewables (Sandalow et al. 2022). In 2021, China also pledged to end investment in coal-fired power plants abroad—this investment had been the major component of the BRI energy engagements (see below)—and to increase support for renewable energy in developing countries.

In 2022, after the Russian invasion of Ukraine, relations between China and the West, as well as with Western allies in the Indo-Pacific region, deteriorated further. This represented another push for the CCP leadership to embrace the move towards self-sufficiency in strategic industrial and technological sectors. However, even China has to address vulnerabilities along its low-carbon supply chains. Domestic demand of critical minerals is expected to increase in the next few years, and the Chinese government is enacting several measures to cope with this new reality, resorting to its traditional direct mobilization of the state's economic might at home and abroad. Moreover, despite China's carbon neutrality pledge, coal continues to be regarded as an essential ingredient for the country's energy security and economic growth. This aspect has been reaffirmed by Chinese policymakers since the end of the strict zero-Covid policy (December 2022) and the reopening of the country. Investments in coal mines and coal power plants have grown in 2022 and 2023, making the achievements of low-carbon objectives more challenging and limiting China's potential to expand its global ambitions on climate policy (Meidan et al. 2023). This was already reflected in Beijing's position at the COP 26 in Glasgow, and it was confirmed at the 2022 COP 27 in Egypt and at the 2023 COP 28 in Dubai, where China maintained its cautious approach to energy transition and the fossil fuel phase-down. At the COP 28, however, China sustained the final agreement that called for 'accelerating efforts towards the phase-down of unabated coal power' and the 'transitioning away from fossil fuels in energy systems' (UNFCC 2023). In the run-up to the COP 28, in a bilateral summit held at Sunnylands in California, China and the US had also jointly committed to 'accelerate renewable energy deployment in their respective economies', and contribute to the G20's goals to 'triple renewable energy capacity globally by 2030' (US Department of State 2023). This deal was part of the relaunch of Beijing–Washington climate talks initiated under the Biden administration (see Chapter 5). Meanwhile, the deployment of wind and solar energy in China has continued at a rapid pace and scale since the war in Ukraine. Moreover, renewable energy and low-carbon technology are now framed as important assets in the wider Chinese global strategy, especially vis-à-vis emerging economies and developing countries in the Global South.

In the following section of this chapter, I illustrate in more detail the emergence and main features of the Chinese green energy state. I go on to explain the drivers and modes in which the Chinese green energy state has 'gone out', shaping the country's green foreign energy policy during the 2010s. A special focus is placed on the green component of the BRI. Finally, in the last section, I focus on the innovations in the Chinese green foreign energy policy and green power strategy that followed the 2020–2021 climate pledges and the 2022 war in Ukraine.

Building the Chinese green energy state

As illustrated above, although China's support for renewables dates back to the 1990s, it was only in the early 2000s that the government made more consistent efforts both to accelerate the deployment of wind (first) and solar (later) installations and to create a low-carbon technology innovation and industrial system. The actual structuring of the Chinese green energy state begun in this period and was consolidated in the following decade, when renewables escalated the agenda of the CCP and the main government institutions, as well as their planning and policymaking activities. This process of institutionalization of renewables into China's political and administrative machinery has evolved in the fashion of a developmental state. Although the country is usually considered the quintessential example of a state-led model of capitalism—specifically a bureaucratic and authoritarian version thereof (Buzan and Lawson 2014)—the literature on the developmental state has clarified that this perspective goes well beyond a simple focus on state interventionism in the economy (e.g., Woo-Cumings 1999; Weiss 1998, 2003; Amsden 2004; Chen and Lees 2016). A developmental state approach involves a steady definition of strategic priorities by the political leadership, including their reflexive adjustment over time, organizational arrangements and policies to guide technological and industrial expansion, and structured links among pilot state agencies and market actors. According to this approach, the government 'picks winners' by strategically sponsoring certain sectors and companies, by offering targeted support through science and technology policy, by mobilizing state funds (and/or state-owned banks), by cultivating leadership in the domestic market and abroad, and by granting national business protection. China's green transition has followed these principles (e.g., Chen and Lees 2016; Genç and Akkemik 2018). However, the Chinese developmental state has taken on specific characteristics (Zhang 2018); it has been embedded in a national context in which the markets are tightly controlled by the government and the CCP, which through their formal bureaucracy and penetration of institutions and civil society, actively shape business decisions of SOEs and private companies and limit public participation in environmental governance (Tunsjø 2013; Chen and Less 2016; Li and Shapiro 2020). Moreover, unlike its classic East Asian predecessors—'developmental bureaucratic states' such as Japan and South Korea (Ó Riain 2004)—the Chinese developmental state has assumed a more fragmented institutional setting. This is characterized by a network of pilot agencies that guide technology and industrial development, rather than a single agency, and by a certain degree of decentralization of economic management in the hands of provincial and local governments (Zhang 2018). This decentralization, however, works within the framework of a top-down political concentration of power in the central institutions of the CCP and the state (Xu 2011).

Institutional fragmentation is particularly evident in the area of renewables. Unlike the fossil fuel sector, centred on a few vertically integrated SOEs that dominate the field, the Chinese green energy sector rests on a multiplicity of state-owned and private companies active in different segments of green business, with further variances from one industry to another (e.g., from solar to wind energy). This setting is matched by a plethora of governmental agencies and CCP organs that deal with a diversity of specific issues concerning renewables and low-carbon technologies. The two most important and interrelated layers of this complex institutional architecture, however, are the one involved in renewables support policies (e.g., feed-in tariff, subsidies, and targets) and the one involved in green energy research, technological development, and the upscaling of manufacturing. Over time, the former layer has been linked with the Chinese climate policy, while the latter has been gradually embedded within the country's wider industrial strategy. This strategy has evolved in an incremental manner in the last decades, although in the recent years it has acquired a more pronounced strategic dimension. Following a brief overview of Chinese energy governance, the evolution of these two layers from the early 2000s is examined in greater detail below. Although the analysis emphasizes wind and solar business and the power sector, it also deals with other significant low-carbon technologies, such as batteries, EVs, and green hydrogen.

Since its dissolution, in 1993, China has not had a unified Ministry of Energy: energy governance relies on inter-ministerial collaboration and negotiation (Andrews-Speed 2012; Andrews-Speed and Zhang 2019; IEA 2022c). Coordination was improved with the government reforms realized at the turn of the millennium, which placed energy governance under the NDRC. This powerful institution works under the State Council—the National People's Congress's chief executive body controlled by the Central Committee of the CCP—and exercises control over long-term planning (including the five-year energy plans), pricing, large investments, and other issues central to energy policymaking. In 2005, the NDRC was also put in charge of the implementation of the CDM. Later, a Department on Climate Change was created within the NDRC, which also become the leading institution responsible for Chinese climate policy. In 2008, the National Energy

Administration (NEA) was created within the NDRC to translate the five-year planning into more specific plans and policies. The NEA regulates several sectors, including coal, oil, gas, electricity, and nuclear power, and supports science and technology research on energy efficiency and renewable energy. Energy research and development (R&D) is also promoted by the Ministry of Science and Technology, which manages the main funds for innovations. Several other ministries, agencies, and SOEs have been mobilized for these activities as well. Provincial and local governments also provide policy and financial support for the implementation of national energy plans and goals. The Ministry of Finance is involved, together with the NDRC and the NEA, in determining the subsidies for renewable energy. In 2010, the National Energy Commission was created within the State Council. This body gathers representatives from all relevant ministries and agencies, such as the NDRC and the NEA, and functions as an umbrella mechanism for high-level coordination of energy policymaking. The coordination of energy issues is also promoted through the CCP organs, particularly the so-called National Leading Small Groups within the State Council. During the 2000s, the Leading Group on Energy was chaired by prime minister Wen Jibao (2003–2013), and its successor—the Leading Group on Climate Change, Energy Saving and Emission Reduction—by the subsequent prime minister, Li Keqiang (2013–2023). In 2021, a Leading Group on Carbon Peaking and Carbon Neutrality was also set up within the NDRC (and chaired by Vice President Han Zheng) to support the implementation of the '1+N' policy framework.

The Chinese power sector was directly in the hands of the government until 2002, when the former Ministry of Electrical Power was corporatized with the separation of transmission and generation and the creation of a small group of state-owned 'energy giants': two large grid companies (State Grid Corporation of China and China Southern Power Grid) and few major power producers. These include the so-called 'big five'—Huaneng Group, Huadian Group, Datang Group, the China Energy Investment Corporation, and State Power Investment Corporation—which own approximately 60% of the country's generation assets, both fossil fuels and renewables (Meidan et al. 2023), and other large SOEs such as the Three Gorges Corporation, and the China General Nuclear Power Group. In 2003, the State-Owned Asset Supervision and Administration Commission (SASAC) was established to improve the government's control over the Chinese SOEs, which nevertheless remain important policy actors with significant resources, authority, and links with the party at central and local levels. Other ministries are also involved in the governance of the energy sector, such as the Ministry of Land

and Resources, the Ministry of Industry, and the Ministry of Environmental Protection. In 2018, a government reform created a new Ministry of Ecology and Environment, which assumed responsibility for environmental issues previously dispersed among a number of other agencies, including those for CO_2 emissions, which were under the NDRC. Figure 3.3 presents a simplified scheme indicating the main actors involved in Chinese energy governance and how they relate to each other.

Since the NDRC's 2003 Wind Power Concession Programme several other measures were implemented to support the deployment of renewables, which were gradually more firmly embedded in the high-level five-year planning process (Table 3.2). In 2006, the first Renewable Energy Law was enacted. This law introduced various mechanisms aimed at accelerating the diffusion of renewables, such as national renewable energy targets, mandatory market shares for renewable energy for large-generating power companies, mandatory access to the grid for wind and solar power installations, a feed-in tariff system modelled on the German experience, and a special fund for renewables. In 2007, the NDRC issued the Medium- and Long-Term Plan for Renewable Energy Development setting more specific targets and mandating larger-power SOEs to produce a (minimum) percentage from (non-hydro) renewable sources. In 2008, renewable energy was formally incorporated into the FYP process, elevating the matter to national strategic status. In the following years, new FYPs on renewable energy with targets and indications

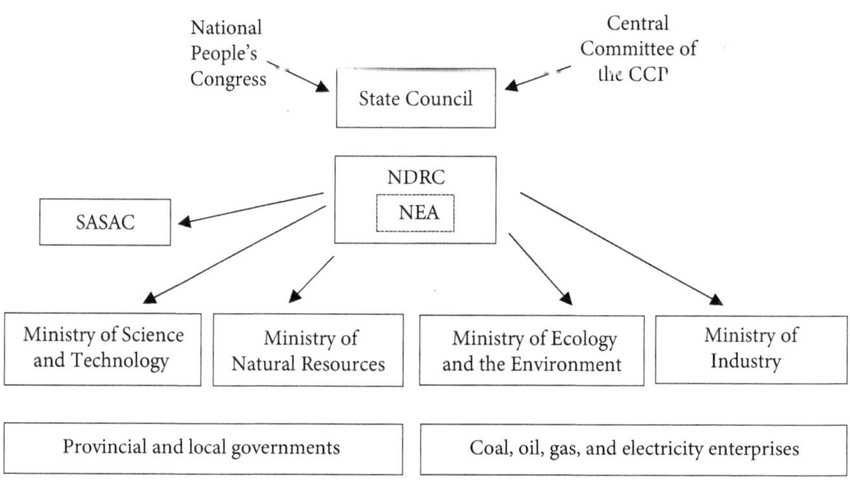

Figure 3.3 The main actors in Chinese energy governance

Source: Author's elaboration from Andrews-Speed and Zhang (2019, 87–88)

were launched, and additional measures to subsidize renewables were implemented, including feed-in tariffs for wind and solar PV (Table 3.2). This subsidy-led development continued until 2015, when a gradual phasing-out process began. In 2016 and 2018, the NDRC announced that China's PV and wind feed-in tariffs would be determined by tendering and auction to further reduce subsidies. This process ran in parallel with the introduction of additional market-based instruments, such as green electricity certificates and the consumption quota mechanisms, or renewable portfolio standard, which oblige primarily power producers and consumers to sell or buy an increasing proportion of their electricity from renewable energy (Table 3.2). In 2021, China also officially launched its national ETS (already announced in 2017). This new tool, covering the power sector and managed by Ministry of Ecology and Environment, will coexist with the pilot carbon markets projects previously launched at the local level. However, its implementation has been problematic, and overall the Chinese emissions trading system has remained in a nascent state of development (Meidan et al. 2023). Also, in 2021, and building on a 2015 similar initiative, the People's Bank of China, the NDRC, and the China Securities Regulatory Commission jointly issued a new version of the Green Bond Endorsed Projects Catalogue, also known as the China Taxonomy, to promote green investments. This initiative resulted from a collaboration with the EU (Cheng and Zhang 2023).

Table 3.2 Main plans and supporting measures for renewable energy generation in China (2003–2021)

Year	Measures/plans
2003	Wind Power Concession Programme
2006	Renewable Energy Law
2007	Medium- and Long-Term Development Plan for Renewable Energy
2008	11th FYP for the Development of Renewable Energy
2009	Feed-in tariff for wind power
2009	Golden Roofs Initiative and Golden Sun Programme
2011	Feed-in tariff for solar PV
2012	12th FYP for the Development of Renewable Energy
2015	Special Fund for Renewable Energy Development
2016	13th Five-Year Plan for the Development of Renewable Energy
2017	Renewable Energy Green Power Certificate Issuance
2019	Renewable Electricity Consumption Quota Mechanism
2021	14th Five-Year Plan for the Development of Renewable Energy
2021	National ETS

Sources: Author's elaboration from Nahm (2021); Zhao et al. (2022); Song, Jia, and Jiao (2022)

Moreover, since the 12th FYP, renewable energy efforts were integrated with the Chinese climate targets (Table 3.3). In both cases, China's ambitions were gradually scaled up. Until the 14th FYP, the government was able to achieve its stated goals. These usually were set in such terms that the accomplishment did not require a drastic departure from existing trends. However, after the 2020 carbon neutrality pledge, the government set more challenging targets in its first (2015) and second (2021) Nationally Determined Contributions (NDC) drafted under the Paris Agreement (Table 3.3).

In connection with these developments, Chinese governmental agents have increased their efforts to promote renewable energy technology and industry by providing a wide range of incentives from the research and innovation stage to manufacturing and commercialization (Hopkins and Li 2016; Lewis 2017; Andrews-Speed and Zhang 2019; Kornes 2020; Nahm 2021). Since the early 2000s, the Ministry of Science and Technology has increased its R&D funding for wind and PV technologies, and promoted the establishment of national-level research centres and laboratories, also in connection with universities, to help raise technical standards and supply technological solutions for manufacturers. This policy was aligned with the broader Chinese innovation and industrial strategy, which aimed at moving away from reliance on foreign technologies and at building up local manufacturing capacity in strategic sectors. This approach was improved in 2006 under the 'indigenous innovation' framework that permeated the 11th FYP period (2006–2010). The goal of the CCP leaders was to upscale manufacturing to boost exports, while at the same time improving the innovation capacity of Chinese companies. This was promoted by the government by introducing a set of measures for promoting technology transfer and the localization of wind and solar technology. Government procurement was leveraged to sustain domestically development technologies, and Chinese scientists and entrepreneurs abroad were also encouraged to return to China. In addition, trade policy was used to protect national business and encourage local manufacturing and industry development. For example, the Ministry of Commerce included renewables in the catalogue of technologies receiving funding and incentives for export. However, cooperation between Chinese companies and foreign companies—especially from Europe and the US—was also important for technological transfer and learning (Nahm 2021). Provincial and local governments granted additional support for upscaling green manufacturing, such as preferential prices for access to land, reduced electricity tariffs, direct investments, and various other subsidies. Although provincial and local activities were not always fully coordinated with directives emanating from Beijing, the combination of the measures

Table 3.3 Main climate and renewable energy targets and achievements in China

Target	12th FYP (2010–2015)	Achieved by 2015	13th FYP (2016–2020)	Achieved by 2020	14th FYP (2021–2025)	2015 NDC	2021 NDC
Climate	17% decrease in carbon intensity from 2010 level	20% decrease in carbon intensity from 2010 level	18% decrease in carbon intensity from 2015 level	18.8% decrease in carbon intensity from 2015 level	18% decrease in carbon intensity from 2020 level	60–65% decrease in carbon intensity from 2005 level by 2030	At least decrease 65% in carbon intensity from 2005 level by 2030
Renewable energy	11.4% non-fossil share of primary energy	12% non-fossil share of primary energy	15% non-fossil share of primary energy	15.9% non-fossil share of primary energy	20% non-fossil share of primary energy	20% non-fossil share of primary energy	25% non-fossil share of primary energy
Wind power	100 GW	129 GW	200 GW	280 GW	—	—	1,200 GW by 2030
Solar power	35 GW	43 GW	150 GW	253 GW	—	—	

Sources: author's elaboration from Lewis (2023a, 29)

enacted at the national and subnational levels were effective in boosting wind and solar industrial capacity. By the end of the 11th FYP period, in particular, China had already become the world's largest manufacturer of PV technologies. There were also differences between the wind and solar sectors. In the former, the domestic market was the main priority, with the demand driven by national subsidies and incentives. Conversely, the growth in the solar PV industry was initially driven by exports, also because there were yet no specific incentives for PV installations and the 2006 Renewable Energy Law was not favourable to their deployment. Moreover, a key role in the wind business was played by the large SOEs, whereas the development of the PV sector was guided by a more diverse group of companies, in some cases private enterprises created by scientists and engineers.

The Chinese green developmental state approach was strengthened by the 12th FYP (2011–2015), when renewable energy, industrial, and innovation policies further converged (Figure 3.4). Renewable energy technologies were included in the 2010 State Council's list of Strategic Emerging Industries, thus opening the way for additional political and financial backing by national and local governments, as well as by the Chinese policy banks. At the same time, solar, wind, and EVs became a key focus area for energy innovation funding and activities—further increased in relation to the previous period—which now targeted the domestic market and manufacturing innovation. After the 2008 global financial crisis—and the reduction of incentives for renewables in several Western countries—the international demand for Chinese PV diminished dramatically. This situation, additionally worsened by the trade disputes with the EU and the US, caused a large crisis for the solar industry now suffering from overcapacity. The government responded to this challenge by sustaining the sector with its post-crisis recovery plan. It also issued new subsidies for the deployment of PV installations (i.e., the Golden Roofs Initiative and Golden Sun Programme) and a national feed-in scheme for solar PV in 2011 (Table 3.2). Moreover, it mobilized its policy banks to provide additional financial support to Chinese PV manufacturers and for the deployment of solar energy. The CDB and CHEXIM also provided support for the wind business, widely contributing to the growth of renewable energy capacity installed in China (Kong and Gallagher 2020; Allen et al. 2021).

During the 13th FYP period (2016–2020), renewables and low-carbon technologies once again figured prominently among the key areas targeted by Chinese innovation and industrial policy (Figure 3.4). In these years, public spending for energy R&D significantly increased, it reached $8.4 billion in

2020 (IEA 2022c). With this rise, in absolute terms, China became the world's second largest energy R&D spender after the US. Moreover, R&D budget increased, especially in the area of renewables and low-carbon technologies; their share within the overall R&D energy budget grew from 35% to nearly 50% in 2020 (IEA 2022c). The 13th FYP was realized after the launch of Made in China 2025. This initiative—which also covered renewable energy manufacturing, batteries, and EVs—aimed at improving the country's technological and industrial capacity in strategic sectors through a combination of subsidies, national champions, and foreign acquisition (Wübbeke et al. 2016). On the one hand, the government wanted to continue reducing its technological dependence on other countries; on the other hand, it aspired to foster novel innovations in priority areas and nurture national companies for global competition. This was the case for EVs and batteries (especially lithium-ion batteries) where the support granted by the state, local governments, SOEs, and private business was instrumental in bringing selected companies (e.g., battery giants such as CATL and BYD) to global industrial scale in a few years (Ladislaw et al. 2021). For this result, an important role was also played by large state-backed venture capital funds, which have been involved in the development of other low-carbon technologies as well (e.g., IEA 2022c).

The trends initiated under the two previous five-year plans continued with the 14th FYP (2021–2025), which was introduced after the Covid-19 crisis, the launch of the 'double circulation' strategy, and in the midst of growing competition between China and the West, especially the US. This evolving international landscape further added a strategic dimension to the Chinese developmental state approach. Moreover, the 14th FYP was introduced after the launch of the China Standard 2035 industrial strategy. This initiative was originally communicated by the Chinese leadership in 2018, although its implementation commenced only in 2021–2022. Ideally, China Standard 2035 represents an evolution of Made in China 2025: it shares the similar goal of fostering the global leadership of the country in strategic industrial sectors. However, it shifts the focus towards standardization at home and abroad—by targeting international standard-setting organizations—as a way to strengthen the Chinese R&D ecosystem, shape the future direction of technological innovation, and reinforce industrial upgrade (Seaman 2020). Low-carbon technology is included in this new industrial strategy, which is in line with China's post-2020 decarbonization goals and also reflected in the innovation priorities highlighted in 14th FYP (Figure 3.4). In this regard, the plan's goal is to maintain Chinese leadership in manufacturing

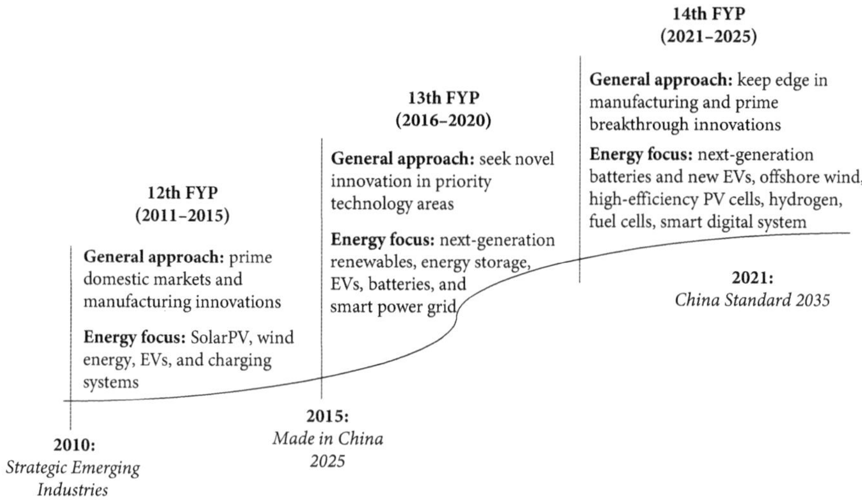

Figure 3.4 The greening of Chinese energy innovation priorities and industrial policy

Notes: Italics denote key policy documents for each period
Sources: Author's elaboration from IEA (2022c) and the main text

and foster breakthrough innovations. In the area of renewables and low-carbon technologies, the focus is on (next generation) batteries, offshore wind, high-efficiency PV, smart grids, and green hydrogen, which is identified as a 'frontier' area for focused advancement in research, manufacturing capacity (especially for electrolysers and fuel cell EVs), and deployment.

This focus on hydrogen was confirmed in 2022, when the NDRC and the NEA issued the first Medium- and Long-term Development Plan for Hydrogen, covering the period 2021–2035. China is already the world's largest producer and consumer of hydrogen, primarily hydrogen derived from fossil fuels. This plan, by contrast, has a stronger emphasis on promoting green hydrogen, with a target of 100,000 to 200,000 tons per year by 2025. It is worth noting, however, that even before the launch of this national strategy, a number of provincial and local governments had formulated hydrogen plans (Gong, Boute, and Quitzow 2023). In this sector, the Chinese approach is mainly inward-looking: the country has vast domestic renewables potential for green hydrogen production and does not need to develop international import routes. Policymakers are therefore encouraging foreign investments and joint ventures—especially with European companies—in order to benefit the transfer of technology and know-how (Quitzow and Gong 2023). Moreover, many Chinese SOEs have also been investing in hydrogen projects. Although the 14th FYP has placed a greater emphasis on a market-oriented approach to research, development, and innovation, paralleling the ongoing shift in the area of renewables support policies, SOEs continue to represent the building blocks of the Chinese green developmental approach and of the country's energy sector. Following the 2020 carbon neutrality pledge, the Chinese leadership has mandated SOEs to be at the forefront of the country's green transition. SOEs have been instructed to build targeted action plans, increase relevant budgets, and carry out R&D in and accelerate deployment of key low-carbon technologies, such as large PV installations, offshore wind, and green hydrogen (IEA 2022c). Finally, critical minerals were also included among the national priorities of the 14th FYP.

As anticipated, China has a strong presence in the supply chains of cobalt, lithium, nickel, and REEs. However, growing global competition over these resources and the anticipated increase in demand to meet China's renewable energy and climate targets has convinced the government to address potential vulnerabilities and reassert a stronger control over the sector. The New Energy Vehicle Industry Development Plan (2021–2035), launched in 2021 by the State Council, provides measures for improving battery recycling, strengthening the development of the value chain of power batteries, and encouraging companies to improve their capacity in the area of lithium, nickel, cobalt, and other key mineral resources. In 2021, the Ministry of Industry, the Ministry of Science and Technology, and the Ministry of Natural Resources also published the 14th FYP for Raw Material Industry Development. This plan reasserted the centrality of REEs for China and initiated a process of consolidation of the sector that will result in the establishment of a large SOE under the SASAC, i.e., the China Rare Earth Group, the world's largest REEs producer.

The Chinese green energy state goes out

The Chinese green energy state has gradually but steadily developed an external reach. This has grown in parallel with its consolidation at home and has expanded the scope of the country's green foreign energy policy. This evolution can be traced in Chinese bilateral and minilateral diplomatic engagements on renewable energy and climate change, in China's growing role in climate finance, and, more significantly, in the areas of industry, trade, and investments where China has deployed its economic tools to increase its influence, especially in the Global South. In line with the BRI, China has also sought to promote its own green developmental model in the Global South, whereas it has largely underplayed the use of regulatory tools.

Chinese bilateral green diplomatic engagements emerged in the late 1990s, but have escalated since the mid-2000s (Figure 3.5). This trend reflects internal developments on the matter and has been guided by institutions such as the NDRC, the NEA, the Ministry of Science and Technology, and the Ministry of the Environment. These actors have signed an increasing number of agreements promoting policy exchanges and cooperation on renewable energy and climate change, primarily with pioneering European countries (e.g., Denmark and Germany), the EU, and the US (Figure 3.5; Table 3.4). Bilateral relationships with the US intensified during the Obama era, including high-level policy dialogue, cooperation in the area of R&D (e.g., with the establishment of the US–China Clean Energy Research Center) and business dialogue, e.g., with the launch of the US–China Renewable Energy Industry Forum (Lewis 2023a). However, this cooperation was halted under the Trump administration (see Chapter 5). Although China's bilateral engagements have increased both in the area of renewable energy and climate change, Chinese policymakers have assigned more importance to the former (Figure 3.5). This element also characterizes other aspects of the country's green foreign energy policy. Moreover, Chinese bilateral green diplomacy has focused primarily on the Global North—with regard to these countries China was a policy taker rather than policy shaper—with a few exceptions relating to other emerging powers such as Brazil and India (Table 3.4).

For engaging the Global South, Beijing has generally used a system of China-led minilateral forums, such as the Forum on China–Africa Cooperation, the China Arab States Cooperation Forum, the China–Community of Latin American and Caribbean States (CELAC) Forum, the Lancang Mekong Cooperation Framework, and the China–ASEAN Environmental Cooperation Forum. Since the 2015 Paris Agreement, this type of forum diplomacy, led by the Ministry of the Environment—and in several cases supported by grants for capacity building, training, and technical cooperation through regional centres (e.g., the China–ASEAN Environmental Cooperation Center, the Lancang–Mekong Environment Cooperation Center, the China–Africa Environmental Cooperation Centre)—has intensified, more clearly incorporating a climate dimension. These forums have also been used to increase dialogue and cooperation in the framework of the BRI. This was particularly the case in South East Asia, through the China–ASEAN Environmental Cooperation Forum, which, since 2015, has been linked to several BRI green initiatives involving a plethora of state actors, business organizations, and NGOs (Liao 2022). Many of these initiatives were also backed by UN agencies, helping China to increase its green profile vis-à-vis the Global

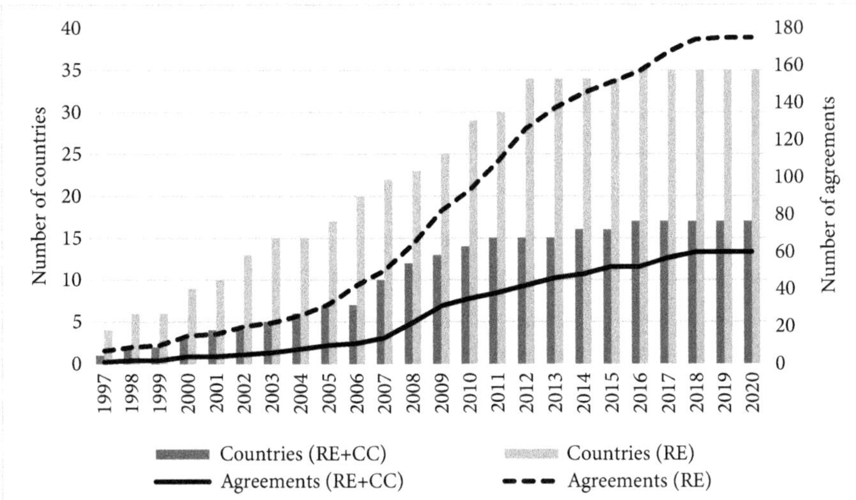

Figure 3.5 Chinese bilateral engagements on renewable energy (RE), and renewable energy and climate change (RE+CC)

Notes: This figure has been compiled by searching the Lewis's database with the entry 'renewable energy', and 'renewable energy AND climate change'
Source: Author's elaboration from Lewis (2023b)

South. Although this Chinese-led (minilateral) forum diplomacy also covers the harmonization of rules and standards, its focus remains on soft forms of policy exchange and cooperation: Chinese platforms emphasize consensus-building and diplomatic functions rather than promoting regulatory regimes or shaping the domestic policy framework for renewable energy in third countries. This approach is in line with the Chinese preferences for limited intrusiveness of international governance and adherence to host countries' rules and standards. Nevertheless, China has used these diplomatic forums, which also include people-to-people dialogue, to promote its soft power in the Global South, in parallel with the growing domestic focus on green development and ecological civilization (e.g., Rauchfleisch and Schäfer 2018; Nedopil 2022).

Since the Paris Agreement, China has also expanded the scale and scope of its multilateral climate finance (e.g., Yeophantong and Goh 2022: 81). Following domestic advances, China has become more active as a provider of green financing and technology transfer to the Global South. In 2016, China pledged to provide $2 billion for the South–South Cooperation Assistance Fund to help meet the post-2015 UN development agenda, which Beijing supported. In the same year, China launched the South–South Climate Change Fund, pledging to commit $3 billion and to build ten low-carbon demonstration projects, one hundred climate change adaptation

Table 3.4 Chinese bilateral agreements on renewable energy (RE), and renewable energy and climate change (RE+CC) with countries in the Global North and Global South

	Number of agreements	
	RE	RE+CC
Global North		
USA	30	9
EU	15	5
European pioneering countries (^)	64	27
Europe (^^)	12	2
Others Global North (*)	22	9
Total Global North	143	52
Global South		
Brazil	12	5
India	7	2
Others Global South (**)	13	1
Total Global South	32	8

Notes: (^) European pioneering countries = Denmark, Finland, France, Germany, Italy, Netherlands, Spain, Sweden, and the UK; (^^) Europe = other EU countries and Norway; (*) Other Global North = Australia, Canada, and Japan; (**) Other Global South = Costa Rica, Cuba, Pakistan, South Africa, Thailand, Saudi Arabia, Venezuela, Mexico, Russia, Turkey, and the UAE
Sources: Author's elaboration from Lewis (2023b)

and mitigation projects, and one thousand training places in developing countries (the so-called '10-100-1000' plan). This latter initiative has been implemented primarily through equipment donation and the organization of training modules, rather than supporting renewable energy projects (Ma 2017; Yeophantong and Goh 2022). Moreover, although the Chinese government reiterated its pledges in subsequent years, problems of coordination and limited political commitment hindered the effectiveness of this plan (Wang 2018). The situation began to improve after 2018, when the China International Development Cooperation Agency (CIDCA) was established as the country's main foreign aid agency under the State Council, a role that was previously played directly by a department within the Ministry of Commerce. However, by the end of 2022, the South–South Climate Change Fund had delivered only approximately $286 million worth of projects; the 10% of the pledged amount and less than 2% of the hundreds of billions of dollars spent in the Global South by Chinese companies and banks for realizing investment projects and opening new trade routes through the BRI (Tsang, Schäpe, and Hackbarth 2023). It is this perspective, focused on trade and investment and commercial diplomacy, that has constituted the main component of Chinese green foreign energy policy.

As anticipated, the first manifestation of this policy can be traced to the 'going-out' strategy in the mineral sector. More broadly, the 'going-out' strategy had a strong energy component. However, this centred on fossil fuels and aimed at increasing China's energy security. With a view to achieving this goal, Beijing has mobilized not only its national oil companies, but also its major policy banks—the CDB and the CHEXIM—which have provided a growing amount of development finance for oil and gas projects around the world. After the 2008 global financial crisis, the CDB and the CHEXIM—which operated in coordination with Chinese state-owned commercial banks and SOEs—became the world's largest overseas public lenders for projects in the power sector, surpassing the Western-led multilateral development banks. However, the vast part of their overseas lending targeted coal-fired power plants and large hydropower projects. From 2000 to 2020, the CDB and the CHEXIM mobilized approximately \$117 billion for power projects worldwide, of which approximately 44% was for coal-fired power, 37% for hydropower, and only approximately 4% for wind and solar energy (Kong and Gallagher 2021). This amount accounted for 6% of global public finance flows for solar and wind energy; by contrast, the European Investment Bank (EIB) was responsible for 24% in the same period (Kong and Gallagher 2021). The Chinese focus on coal and hydropower reflected the drivers of its domestic energy sector. In the late 2000s, coal and dam construction companies were suffering from overcapacity as a result of a slowdown in electricity demand. China's policy banks were consequently mobilized to support their expansion abroad. However, China's policy banks (and companies) also filled a demand in developing countries, as traditional donors were moving away from coal projects and large dams. Moreover, unlike their Western counterparts, Chinese policy banks did not apply policy conditionalities, an element that worked as a further incentive for several countries in the Global South. This type of Chinese foreign energy engagement further increased after the launch of the BRI, as China's policy banks were mandated to support its industrial and infrastructure objectives. In Africa, Chinese companies became the dominant provider of power generation, as well as transmission and distribution technologies (e.g., IEA 2016). Overseas energy financing by the CDB and the CHEXIM, however, decreased in the late 2010s and was to be drastically reduced after the Covid-19 crisis (Ma, Han Springer and Shao 2022). In 2021–2022, China's policy banks provided no new energy finance commitments to foreign governments for the first time since 2000 (CGEF 2022; 2023).

It was only in the early 2010s that Chinese policy banks begun supporting the internationalization of the wind and solar industry. Previously, this process had been guided by PV and wind turbine manufactures. These companies wanted to integrated their business downstream, investing in renewable energy projects in Europe and North America, or to gain access to new technology, such as in the case of the Goldwind's acquisition of 70% of the German wind turbine company Vensy in 2008 (Tan et al. 2013; Curran, Lv and Spigarelli 2017; Andrew-Speed and Zhang 2019). In 2011–2012, however, the combination of excess manufacturing capacity with decreasing international demand meant that China's policy banks not only had to sustain the wind and solar industry at home but also assist their expansion abroad. Nevertheless, their engagement in this field remained very limited in comparison to the coal and hydropower sector. Moreover, the Chinese approach in this area took different forms: a market-seeking and technology-seeking perspective in the Global North and a broader diplomatic and political engagement in the Global South (Kong and Gallagher 2020, 2021). In Europe, for example—particularly in Italy (2010), Romania (2013), and Bulgaria (2013)—the CDB provided financial support to Chinese PV manufactures willing to expand their operations through mergers and acquisitions and attracted by the generous support schemes provided in these EU countries for the deployment of solar installations. The CDB had already established close connections with these manufacturers at home, and teamed up with them to expand their domestic collaboration abroad. In addition, the CDB provided financing to Chinese manufacturers for acquiring new technology, such as in the case of Hanergy, which was able to purchase several US and European companies specializing in thin-film solar technology between 2013 and 2015 (US Trade Representative 2018). The first type of operations, however, encountered several problems, and since then, Chinese policy banks have not financed similar ventures. By contrast, both the US and European countries became more protectionist in the area of renewable energy from the mid-2010s, halting Chinese acquisition of key low-carbon technologies.

In the Global South, China's policy banks targeted Pakistan, Ethiopia, Argentina, Lesotho, Cameroon, and Kenya (Figure 3.6). In the latter five cases, national governments were the actual borrowers of CDB and CHEMIX loans, and cooperation with China took place under government-to-government frameworks aimed at promoting wider economic and trade relations (Kong and Gallagher 2020). In the case of the wind and solar projects in Pakistan, Chinese assistance focused on project financing, although the loans from Chinese policy banks were provided under an agreement between Beijing and Islamabad. This overseas development finance granted by the CDB and the CHEXIM entailed Chinese firms constructing the projects and

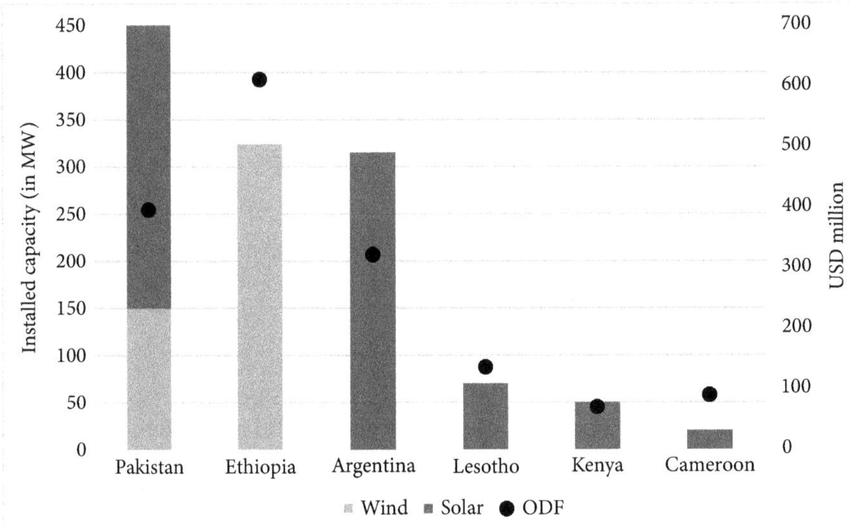

Figure 3.6 Renewable energy projects promoted by China's policy banks in the Global South (2011–2020): installed capacity and overseas development finance (ODF) provided

Notes: As of 2020, other smaller projects were under consideration in Togo, DRC, and Sudan (solar), and Ecuador (wind)

Sources: Author's compilation from Kong and Gallagher (2021) and CGEF (2022)

supplying equipment such as wind turbines and solar PV (Kong and Gallagher 2020). In Cameroon, the project was developed by the Chinese SOE Sinohydro, as part of a larger hydropower installation. However, together with this component aimed at increasing Chinese exports, the government used renewables to improve bilateral relationships with these countries, thus supporting their own national developmental agenda. This was the case especially with Pakistan, Ethiopia, and Argentina. In the case of Argentina, which officially joined the BRI in 2022, China was upgrading its relationship as part of a process of increasing involvement in Latin America. Ethiopia is one of China's strongest African partners and holds a strategic position on the Horn of Africa. Pakistan, through the China–Pakistan Economic Corridor is a key partner and enabler of the BRI. China also invested in the expansion of the national grid in Pakistan, promoting high-voltage direct current (HVDC) lines.

To support electricity grids expansion, China launched the Global Energy Interconnection (GEI) initiative in 2015. Strongly backed by the State Grid Corporation of China—the main developer of HVDC technology in China—this initiative envisaged the creation of a global network of transnational electricity interconnectors to enable the energy transition. The GEI represents an extension of the BRI into the electricity grid sphere (Downie 2020). However, this initiative is a vision rather a specific plan for the implementation of a connectivity strategy. It serves more as demonstration of the technical potential of Chinese technology and as a vehicle for advancing the country's own industrial policy objectives. More concrete efforts relating to transnational electricity networks have been pursued by China in South East Asia, where Beijing has engaged with the ASEAN and the Greater Mekong Subregion Forum (this forum brings together China, Cambodia, Lao, Myanmar, Thailand, and Vietnam). In these regions, too, China has not sought to promote broad regulatory frameworks; rather, it has advanced grids' connectivity on a bilateral basis and with the involvement of its SOEs (Feng et al. 2020; Westphal, Pastukhova, and Pepe 2022). Since the lunch of the BRI, these companies have increased their investments in the electricity grid infrastructure of South East Asian countries (for example, the State Grid Corporation holds a 40% stake in the Philippines' national electricity transmission company). The development of transnational interconnections has nevertheless remained problematic and very limited (China has electricity interconnections with Laos, Myanmar, and Vietnam).

In the same period, Chinese SOEs have increased overseas investments in the power transmission and distribution business beyond South East Asia (Henares and Delina 2022). The State Grid Corporation and China Southern Power Grid have acquired minority stakes in European grid companies (in Portugal, Italy, Greece, and Luxemburg), as well as in Australia. In Latin America, the State Grid Corporation, the China Southern Power Grid, and the Three Gorges Corporation entered in the Brazilian, Chilean, and Peruvian electricity grid sector, often replacing European and US companies. Unlike in Europe and Australia, where investments were guided by the desire to seek new markets, in Latin America the expansion of the Chinese SOEs was paralleled with growing bilateral political relations. Brazil, a BRICS member, concentrates the largest part of State Grid's assets outside China and has played an important role in the global push of Chinese companies (Batista Barbosa 2020). Moreover, both Chile and Peru have joined the BRI and consolidated their relations with Beijing.

Within the context of the BRI, the foreign direct investments of Chinese SOEs, especially those active in the power sector, and the financial support granted by Beijing's major state-owned commercial banks—such as the Industrial and Commercial Bank of China and the Bank of China—have also been instrumental for promoting renewable energy projects. Chinese SOEs, in particular, have come to play a crucial role in this area. Indeed, although

Chinese private wind and solar manufacturers continued to expand abroad during the 2010s, the government mandated its large power companies, in particular, to take over the green BRI investments.

Energy is the first BRI sector in terms of investments, followed by transportation and mining activities (Nedopil 2023a). However, the green component of the BRI, excluding hydropower, remained minimal until 2016–2017 (Figure 3.7). After this period, several initiatives were taken with a view to greening the BRI (see below), and the proportion of investments in wind and solar projects has increased: from 11% in 2017 to an average of 21% in the 2019–2022 period (Figure 3.7). However, this relative growth occurred in the context of a general reduction in BRI overall energy engagement, which averaged approximately $35 billion per year between 2013 and 2017, $30 billion in 2018 and 2019, and only approximately $25 billion between 2020 and 2022 (Figure 3.7).

This translated into an irregular trend of wind and solar capacity being installed in the BRI countries thanks to Chinese companies' foreign direct investments (Figure 3.8). This type of engagement has certainly increased significantly since the commencement of the BRI. From 2000 to 2012, Chinese companies (especially manufacturers such as Goldwind, Hareon Solar, and Chint) promoted around 550 megawatt (MW) of new wind and solar

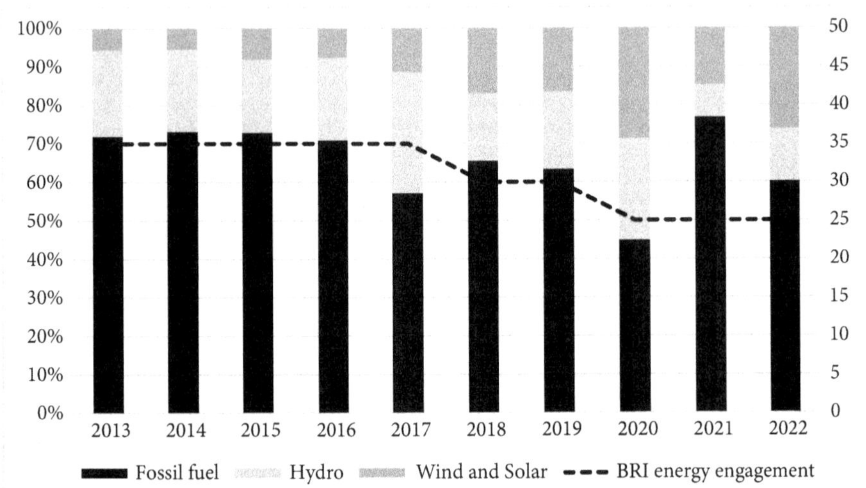

Figure 3.7 China's BRI energy engagement through investment and construction: percentage for different energy sources (left axis) and average amount (in USD billion, right axis)

Sources: Author's elaboration from Nedopil (2023a)

installed capacity, mainly in Europe and Australia (CGPD 2022). On the other hand, from 2013 to 2022, this capacity amounted to nearly 8,200 MW: 3,386 MW in the BRI countries and 4,830 MW in non-BRI countries (Figure 3.8). This data reflects the fact that many Chinese companies targeted more mature and profitable markets for renewable installations in the Global North. In addition, important countries in the Global South that attracted large amount of Chinese green investments are outside the BRI framework. These include Brazil and Mexico, the largest recipients of investments in Latin and Central America. Again, however, Chinese engagement in the Global North was guided primarily by a market-seeking perspective, which also included an important role for mergers and acquisitions (M&A; Table 3.5). In the Global South, by contrast, China has deployed its economic might (also) to cement political relations and promote its developmental state approach along its industrial outreach. As anticipated, however, Chinese engagement in both these areas has been nurtured primarily by its largest power sector's SOEs rather than private (often manufacturer) enterprises (Table 3.5).

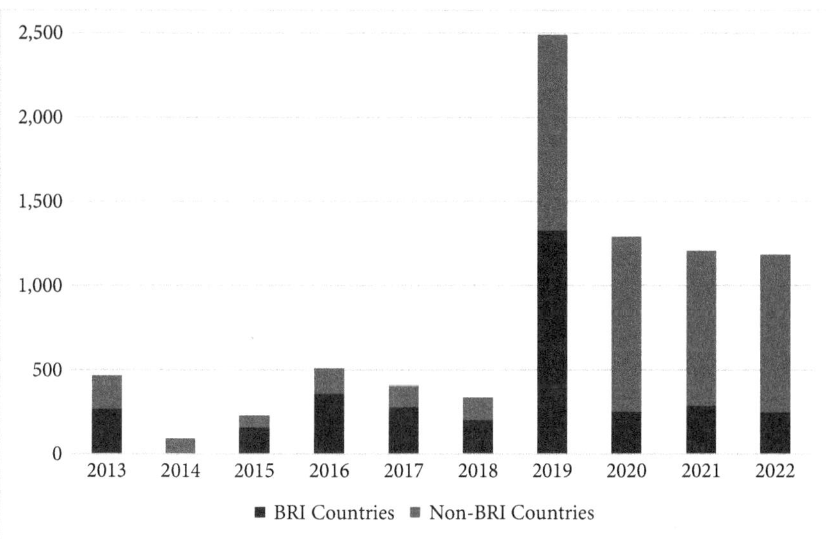

Figure 3.8 Wind and solar projects (in MW of installed capacity) promoted through foreign direct investments by Chinese companies in the BRI and non-BRI countries (2013–2022)

Notes: BRI countries include countries that have signed a BRI-related MoU with China between 2013 and 2022 (data from: https://greenfdc.org/countries-of-the-belt-and-road-initiative-bri/, accessed 22 July 2023)

Sources: Author's elaboration based on CGPD (2022). This data includes only the foreign direct investments by Chinese companies for greenfield projects (i.e. built from the ground up)

Table 3.5 Wind and solar PV projects (in MW of installed capacity) promoted through Chinese overseas investments (2013–2022)

Region	SOEs	Policy Banks	Private Enterprises (PEs)	Total (% SOEs) [% Policy Banks]	Main investing companies
Latin and Central America					
Wind	2710	0	526	3236 (84%)	*SOEs*: China General Nuclear Power Group (§), State Grid Corporation (§), State Power Investment Corporation (§), China Three Gorges Corporation (§), Guodian Group (§)*PEs*: Goldwind (°), Envision Energy (°), Jinko Solar (°), Trina Solar (°), Risen Energy (°)
Solar	1147	315	803	2265 (50%) [14%]	
Total	(% M&A)			5501 (64%)	
Europe					
Wind	2919	0	54	2974 (98%)	*SOEs*: China Three Gorges Corporation (§), China General Nuclear Power Group (§), China Huadian Corporation (§), State Power Investment Corporation (§)*PEs*: Chint (°), Panda Green Energy (ç), Hareon Solar (°)
Solar	825	0	320 (*)	1145 (72%)	
Total	(% M&A)			4119 (57%)	
Asia					
Wind	348	0	0	348 (100%)	*SOEs*: China Three Gorges Corporation (§), PowerChina (^), State Power Investment Corporation (§), China General Nuclear Power Group (§)*PEs*: Risen Energy (°), Suntech (°)
Solar	550	300	234	1084 (50%) [28%]	
Total	(% M&A)			1432 (0%)	

Australia					
Wind	544	0	1319	1863 (29%)	*SOEs*: State Power Investment Corporation (§), China General Nuclear Power Group (§), China Three Gorges Corporation (§)*PEs*: Goldwind (°), Risen Energy (°), Clenergy (°)
Solar	300	0	399	699 (43%)	
Total	(% M&A)			2562 (32%)	
Africa					
Wind	244	324	0	568 (43%) [57%]	
Solar	0	120	22	142 (84%)	*SOEs*: Guodian Group (§), PowerChina (^)*PEs*: Chint (°)
Total	(% M&A)			710 (0)	
North America					
Wind	80	0	0	80 (100%)	
Solar	0	0	35	35	*SOEs*: Guodian Group (§)*PEs*: ReneSola (°), LDK Solar (°)
Total	(% M&A)			115 (0)	

Notes: Europe = EU member states, plus Ukraine, and the UK; Latin and Central America = Argentina, Brazil, Chile, Colombia, Ecuador, Mexico, and Peru; Africa = Cameron, Ethiopia, Ghana, Kenya, Lesotho, and South Africa; North America = US and Canada; Asia = Bangladesh, Kazakhstan, Malaysia, Pakistan, and the Philippines. (*) = This data includes the MW supported by policy banks that provided finance to PEs. (§) = power company; (°) = manufacturer; (ç)= project developer; (^) = construction company
Sources: Author's elaboration from CGPD (2022). This data includes foreign direct investments by Chinese companies for greenfield projects and merger and acquisition (M&A), plus the projects supported by Chinese policy banks.

These SOEs, such as the China Three Gorges Corporation, the China General Nuclear Power Group, the State Grid Corporation, the State Power Investment Corporation, and the Guodian Group, have 'deep pockets' and can support large strategic investments. Their involvement abroad also helps the export of Chinese wind and solar equipment and the diffusion of Chinese technological standards (see below). In Latin and Central America, the first region for renewable energy projects (in terms of installed capacity) supported through Chinese capital, 84% of wind installations have been promoted by SOEs. This number is lower for the solar sector (50%), although another 14% of installed solar capacity has been backed by Chinese policy banks (i.e., the solar project in Argentina). In Europe—the second region for Chinese-supported renewable installed capacity—SOEs have similarly been responsible for the largest part of investments, in both the wind and solar sectors (Table 3.5). Moreover, as in Latin and Central America, many Chinese investments also took the form of M&A (Table 3.5). It is worth noting, however, that China's green engagement has been minimal in the Western Balkans (the exception is a wind farm in Montenegro; CGPD 2022). Chinese economic and diplomatic involvement have increased in this region since the early 2010s. North Macedonia joined the BRI in 2013, followed by Serbia (2015), and by Albania, Bosnia and Herzegovina, and Montenegro (2017). Beijing has also established close political and economic relations with Serbia (Stanicek and Tarpova 2022). However, most Chinese investment in the Western Balkans has focused on transport infrastructure (e.g., railways, highways, and ports) and fossil fuel projects (ibid.). In Africa and Asia, together with SOEs—as showed—Chinese policy banks provided an important contribution to renewable energy projects (Table 3.5). In Australia, SOEs have played a smaller but still important role, especially in the solar sector. It is only in North America that Chinese private manufacturers have been responsible for all the projects in the solar sector; however, the overall Chinese contribution to such projects in this region has been negligible (Table 3.5).

As mentioned, this external reach of Chinese companies also helped Beijing's exports, which, however, has continued to be sustained by the cost-competitive position of Chinese green technology. Europe, Latin and Central America, Asia, and Australia have become the main regions for Chinese solar panels exports (Ember 2023); and a similar pattern characterizes Chinese wind turbine exports, although in this case, North America is also an important destination (US International Trade Commission 2022). Moreover, after the launch of the BRI, China has also increased its green engagement in the Middle East. In 2019, through the state-owned Silk and Road Fund—established in 2014—China acquired a 49% stake in Saudi Arabia's ACWA

Power Renewable Energy Holding, which owns several renewable energy projects across the UAE, South Africa, Jordan, Morocco, and Egypt. In the same year, Jinko Solar made a large investment for the development of solar projects in the UAE. These moves were also important to sustain the export of Chinese green technology, especially PV panels, in the region. In the case of Egypt—a country strategically tied to the BRI—Chinese SOEs participated in the development of the Benban Solar Park, which also received $210 million from the Chinese-led multilateral development bank in 2017, that is, the Asian Infrastructure Investment Bank (AIIB; Hoffmann and Ergenc 2023). However, the AIIB and the New Development Bank (NDB)—the BRICS bank in which China is an equal shareholder, along with the other funding members—as well as the Western-led multilateral development banks, play a negligible role in financing the BRI. Their involvement was estimated to account for less than 1% of all BRI projects as at 2018 (ICBC 2018). As BRI investments are driven by Chinese financial institutions and companies, the greening of the BRI is closely linked to Chinese efforts in this direction.

The greening of the BRI received a boost after 2015, in parallel with the changing Chinese position on global climate policy and amid growing criticism from the international community on the BRI's negative environmental impacts. Criticism of BRI projects also came from academia, think tanks, and environmental policy groups within China itself, thus reflecting the increasing domestic attention these issues receive. However, the emergence of the BRI's green agenda was not merely a response to these criticisms: it denoted a component of the growing external reach of the Chinese green developmental state (Liao 2021). Since 2015, the Chinese authorities have issued several policy documents and regulatory measures for greening the BRI (Table 3.6). This shift was accompanied by a growing discursive focus on green and sustainable development by Chinese leaders—including Xi Jinping—during the first (2017) and second (2019) Belt and Road Forum. In 2019, China launched the International Coalition for Green Development on the Belt and Road as a multi-stakeholder platform comprising environmental ministries, intergovernmental organizations, NGOs, experts, and business, for policy dialogue, communication, and technology exchange on environmental protection, renewable energy, and energy efficiency. However, the implementation of this shift proved to be very difficult. Chinese regulatory measures—both those specifically targeting the BRI and those related to the BRI as they targeted all Chinese overseas investments (Table 3.6)—provided only for voluntary actions by financial institutions and companies. China's main governmental agencies (e.g., NDRC, the SASAC, the Ministry

Table 3.6 Key policies and regulatory measures for greening the BRI (2013–2020)

Year	Initiative	Focus/type of measure	Issued by
2013	Guidelines on Environmental Protection for Overseas Investment and Cooperation	BRI-related/regulatory	Ministry of Commerce (MOFCOM), Ministry of Environmental Protection (MEP)
2015	Visions and Actions on Jointly Building Silk Road Economic Belt and 21st Century Maritime Silk Road	BRI specific/policy document	NDRC, Ministry of Foreign Affairs (MOFA), MOFCOM
2016	Guidelines for Establishing the Green Financial System	BRI-related/regulatory	NDRC, MOFA, MEP, People's Bank of China, China Banking Regulatory Commission
2017	Guidance on Promoting a Green Belt and Road	BRI-specific/regulatory	NDRC, MOFA, MOFCOM, MEP
2017	The Belt and Road Ecological and Environmental Cooperation Plan	BRI-specific/policy document	MEP
2017	Measures for the Administration of Overseas Investment of Enterprises	BRI-related/regulatory	NDRC
2017	Visions and Actions on Energy Cooperation in Jointly Building Silk Road Economic Belt and 21st Century Maritime Silk Road	BRI-specific/policy document	NDRC, NEA
2019	Green Investment Principles	BRI-specific/regulatory	Chinese and international financial institutions (*)

Notes: (*) = This initiative has been jointly developed by the Green Finance Committee of China Society for Finance and Banking and the City of London Corporation's Green Finance Initiative and signed by all major Chinese banks as well as several international financial institutions
Sources: Author's elaboration from Liao (2021), Coenen et al. (2021), and Nedopil (2022)

of Commerce, and the NEA) continued to prioritize investments and industrial goals rather than environmental and climate objectives (Gallagher and Qi 2018; Liao 2021). At the end of 2020, the most explicit environmental requirement for Chinese companies remained compliance with host country rules, which was commonly seen as a low bar in ensuring the environmental standard of BRI projects in the Global South (Voituriez, Yao, and Larsen 2019). Moreover, the green guidelines for financial institutions do not apply to China's policy banks, which are under the direct supervision of the State Council and aim at promoting China's economic interests. As China is not part of the OECD, its policy banks do not apply either the environmental or social standards that their counterparts in the OECD countries do. For example, many of the projects financed by the CDB and the CHEXIM in Asia and Africa are located in sensitive territories, such as national protected areas, critical habits, or indigenous people's land (Yang et al. 2021). It was only after 2020–2021 that Chinese regulatory efforts for greening the BRI and Chinese overseas investments were improved (see below).

Greening the Chinese global strategy

Although the 2020 climate neutrality pledge represented an important step in the Chinese internal and international green energy policy, it was not a path-breaking moment that reoriented Beijing's approach on the matter. In the wake of the Covid-19 crisis and the war in Ukraine, which negatively affected international oil and gas markets, the role of coal as a key component of the country's energy security and economic development was reasserted. The 2021 COP26 in Glasgow was the first to explicitly mention coal and fossil fuel subsidies. However, on the final day of the negotiations, China—together with India and on the behalf of several developing countries—watered down the provisions relating to coal: from 'phase-out' to 'phase-down'. At the 2022 COP 27 in Sharm El-Sheikh, as well as at the 2023 G20 summit in India and at the 2023 COP 28 in Dubai, China reiterated this more conservative approach to energy transition. It also confirmed its efforts to maintain its identity as a developing country, asking for further action (and funding), especially from the Global North. However, the greening of the Chinese global strategy continued after 2020. I am referring here to the process of incorporating green foreign energy policy objectives into the wider Chinese approach to extend its influence in the international system, challenging the US (and the West) leadership at the global and regional levels (e.g., Doshi 2021).

Beijing continued its bilateral and minilateral diplomatic activities in the area of renewables and climate change. In 2021, a new Declaration on China–Africa Cooperation on Combating Climate Change was issued at the Forum on China–Africa Cooperation. It proposed the establishment of a high-level forum on climate change and launched a three-year action plan that provided for additional Chinese investment in low-carbon projects. At the 2021 China–CELAC Forum, China called for increasing cooperation on climate change, energy transition, and low-carbon industries. Then, in 2022,

China established a new China–Pacific Island Countries Climate Action Cooperation Centre to carry out demonstration projects and share Beijing's expertise on low-carbon development. In the same year, renewables and green infrastructures were identified as priority areas of the Plan for Action for Implementing the China–ASEAN Strategic Partnership for Peace and Prosperity (2021–2025). In addition, a China–ASEAN Joint Statement on Strengthening Common and Sustainable Development was issued in 2022, reaffirming the importance of cooperating for the energy transition in the region. It is worth stressing that the Regional Comprehensive Economic Partnership Agreement (RCEP) also entered into force in 2022. This trade deal, agreed to in November 2020 during the Covid-19 crisis, had emerged within the context of ASEAN a decade earlier, but it was also backed by China, which signed it together with the ASEAN members, Japan, South Korea, Australia, and New Zealand. Although this deal does not contain provisions on environmental protection or climate change or on limiting the role of SOEs, it can potentially contribute to further increasing regional exchanges in certain renewable energy technologies by reducing trade barriers and improving technical cooperation (Limenta 2023). Moreover, this regional agreement emerged as a China-backed alternative to the US-led Trans-Pacific Partnership abandoned by the Trump administration (see Chapter 5). During the Covid-19 crisis in 2020, China had also reached a political agreement with the EU in the form of the EU–China Comprehensive Agreement on Investment (CAI); however, this deal was not to be finalized either (see Chapter 4).

Together with the continuation of its minilateral diplomatic activities, China has included a green dimension in all its foreign policy engagements. In 2021, the Shanghai Cooperation Organisation (SCO) established a mechanism of meetings for energy ministers for policy exchanges on energy security, as well as the energy transition and the use of renewables (SCO 2021). At the SCO summit held in Tajikistan in the same year, China launched the SCO–Green Belt Programme to foster the implementation of projects on environmental protection, energy savings, and renewables. In 2022, under Chinese leadership, the SCO (2022a, 2022b) reiterated its goals of increasing cooperation in the area of green and low-carbon technology (and climate change) as well as for improving the reliability and diversification of supply chains for the green economy (this latter issue was repeated at the 2023 SCO summit). At the 2022 BRICS summit, held in China, cooperation on energy transition and low-carbon technology figured prominently on the agenda. Beijing also confirmed its support for the BRICS Plus initiative—already launched by China in 2017—to expand the group to new members from the Global South, including important countries for emerging low-carbon technology supply chains, such as the resource-rich Argentina (lithium) and Indonesia (nickel). At the subsequent 2023 BRICS summit, held in Johannesburg, tackling climate change and developing partnerships for an 'equitable just transition' were the first theme and priorities of the agenda (BRICS 2023). On this occasion, the BRICS leaders finally decided to formally invite six more countries to join the group (from January 2024): Argentina, Ethiopia, Iran, Saudi Arabia, Egypt, and the UAE. After the election of the new President Javier Milei, on November 2023, Argentina, however, reversed its position, choosing to not join the BRICS.

In 2021, the NDB had already expanded its membership with the accession of Bangladesh, Egypt, the UAE, and Uruguay. Moreover, the NDB investment strategy for 2022–2026 provided for dedicating 40% of its funding—expected to total \$30 billion—to projects contributing to climate change mitigation and adaptation and supporting member countries' commitments under the Paris Agreement (NDB 2022). Also in 2021, the AIIB updated its Environmental and Social Framework, requiring projects to be in accordance with the aims of the Paris Agreement and the members' nationally determined contributions (AIIB 2021). It further announced a target of 50% for financing approvals in relations to climate finance by 2025. This commitment was important, as in comparison to the Western-led multilateral development banks, the AIIB had formerly ranked lowest in absolute climate finance volume (Nedopil 2022). In 2022, the AIIB (2022) further reviewed its Environmental and Social Framework to exclude the financing of new coal projects. Moreover, in 2023 it signed a Memorandum of Understanding with the CIDCA to collaborate in addressing climate change and realizing the 2030 Agenda for Sustainable Development. In October 2023, both the NDB and the AIIB strengthened their cooperation with other multilateral development banks, including Western-led banks such as the EIB, the European Bank for Reconstruction and Development (EBRD), and the World Bank, to boost join action on climate change (Reuters 2023a). Moreover, although China resisted the call by the EU and the US to increase its multilateral climate finance and take part in the main UNFCCC mechanisms alongside developed nations, Beijing pledged a further \$1 billion for South–South climate cooperation under a fund renamed the Global Development and South–South Cooperation Fund. This move was part of the Global Development Initiative, launched by Xi Jinping in 2021 to boost sustainable development projects and capacity building in the Global South. Although the actual scope and possible impact of this new initiative are not yet clear, unlike the BRI, it placed emphasis on delivering the UN Sustainable

Table 3.7 Main post-2020 measures for greening the BRI and Chinese foreign investments

Year	Initiative	Focus/type of measure	Issued by
2021	Green Development Guidelines for Overseas Investment and Cooperation	BRI-related/policy document	MOFCOM, Ministry of Ecology and Environment (MEE)
2022	Guidelines for Ecological and Environmental Protection of Foreign Investment Cooperation and Construction Projects	BRI-related/policy document	MOFCOM, MEE
2022	Opinions on Jointly Promoting Green Development of the Belt and Road	BRI-specific/policy document	NDRC, MOFA, MEE, MOFCOM

Source: Author's elaboration from Sandalow et al. 2022 and the Belt and Road Initiative Policies Database, available at: https://greenfdc.org/bri-policies/ (accessed 22 May 2023)

Development Goals, rather than large infrastructural investments, and on closer cooperation with the UN agencies. As part of the Global Development Initiative, Beijing also launched a Global Clean Energy Cooperation Partnership in 2022, to promote investments in renewable energy abroad. However, this is another case in relation to which there is no clarity on the scope of and funding available for the initiative. As in the previous period, bilateral engagements and the investments of Chinese companies continue to be the main component of Beijing's green foreign energy policy and green power strategy. This in a context of a general reduction in Chinese global overseas investments caused by a range of external and internal factors; from China's zero-Covid policy to rising global risks following Russia's invasion of Ukraine and slower economic growth (Kratz et al. 2023).

As highlighted, the greening of Chinese overseas investments was improved after 2020. This process continued in the following years, when additional provisions for greening the BRI and Chinese foreign investments were introduced (Table 3.7). These new initiatives did not specify enforcement mechanisms, and the question of their effective implementation remains open (Sandalow et al. 2022). However, they reinforced the trend initiated in the previous period.

The 2021 guidelines, in particular, were the first ever policy document asking Chinese companies to follow international rules and standards in their overseas activities, departing from the host country principle. The 2022 guidelines, then, underlined that the greening of the BRI is an 'inherent requirement of implementing the concept of green development and advancing the construction of an ecological civilization' (Sandalow et al. 2022, 221). These guidelines also stated that China's investments abroad should aim to facilitate the full implementation of the UNFCCC and Paris Agreement. They further encouraged Chinese companies operating abroad to use international principles and standards or higher Chinese standards when host country standards are absent or too loose (ibid.). This shift might help China to become a rule exporter through its overseas investments. Finally, the 2022 BRI-specific opinions contained detailed provisions for certain sectors (e.g., infrastructure, energy, and transport), setting green targets to be achieved by 2025 and 2030. Moreover, these opinions offered an official operationalization of China's 2021 pledge to stop building new coal-fired power projects abroad: they stipulated an end to completely new projects, but left an ample grey area, such as for the expansion of existing and planned projects (Sandalow et al. 2022). From the making of the pledge in 2021 to the end of 2022, nearly 12.8 gigawatt (GW) of China-backed overseas coal projects (fifteen power plants) were shelved or cancelled following policy changes in host countries and/or withdrawal of Chinese companies (CREA 2022). However, other planned projects have not been halted, including projects of strategic importance for Beijing, such as two coal-fired power plants in Indonesia linked to major BRI industrial facilities promoted by Chinese companies for nickel and steel production (CREA 2022).

At the same time, the greening of the BRI continued. In the first half of 2023, about 41% of BRI energy engagements targeted solar and wind projects, the highest mark since its inception (Nedopil 2023b). This, however, in a context of steady reduction of overall BRI energy engagements—approximately $12 billion in the first half of 2023—also caused by a drop in natural gas- and oil-related projects (ibid.).

Backed by the government, hence, Chinese companies have continued to invest in green projects abroad, including in new fields such as EVs and green hydrogen. Importantly, China has structurally embedded renewables and low-carbon technology into its approach for consolidating bilateral political relations with strategically important countries and regions of the Global South and/or along the BRI. Moreover, despite its (primarily) inward-looking policy in this field, and following its traditional approach, Beijing has mobilized its SOEs to increase its role the nascent international green hydrogen business (Quitzow and Gong 2023). In 2023, China and Brazil (and their respective companies) signed several deals for promoting renewable energy projects and EVs facilities in Brazil, and for improving technological and industrial cooperation. Beijing and Brasilia also discussed the possibility of creating a green investment fund to finance the development of green industry and renewable energy in both countries (Paraguassu and Boadle 2023). Meanwhile, Chinese SOEs (i.e., the State Power Investment Corporation and the Three Gorges Corporation) invested in research and development activities in the Brazilian green hydrogen sector. Chinese SOEs also began cooperating with Pakistan with a view to developing the first green hydrogen projects in that country. At the China–South Africa New Energy Investment and Cooperation Conference held in Johannesburg in 2023, Beijing offered to provide 66 GW of solar power generation to help South Africa stabilize its energy grid and address electricity shortages. Renewables and low-carbon technology also became a structural component of Beijing's rising role in the Middle East, which was well illustrated by the 2023 Saudi Arabia–Iran deal brokered by China. This region is indeed crucial for Beijing's oil supply, and this also explains why countries such as Saudi Arabia, Iran, and the UAE were invited to join BRICS. Nearly half of China's oil imports come from the Middle East; and although Beijing increased its purchases of Russian oil following the war in Ukraine, Saudi Arabia remains its leading supplier. However, in this case as well, China's green engagements were upgraded. In 2022, the first summit between China and the Gulf Cooperation Council (as a group) was held in Riyad. This summit resulted in a Five-Year Action Plan with a strong focus on renewables and low-carbon technology. At the same time, China continued to foster investments in large utility-scale solar projects through ACWA Power, especially in Saudi Arabia and the UAE, supporting their energy transition efforts while sustaining the export of Chinese green technology (e.g., Carpenter 2023). From 2022 to 2023, for example, solar panel exports from China to the Middle East grew by more than 60% (EMBER 2023). In 2023, Saudi Arabia also signed a $5.6 billion deal with the Chinese (private) company Human Horizons for the development, manufacture, and sale of EVs, together with a deal with the state-owned China National Geological and Mining Corporation for expanding copper mining in the country (Aljazeera 2023). Moreover, Riyad and Beijing signed a Memorandum of Understanding (in December 2022) for cooperation in the area of hydrogen. In Egypt, the Chinese SOE China Energy Engineering Corporation—which was already active in the country's renewable energy sector—signed (in 2023) a deal with companies and authorities in Cairo for developing a large green hydrogen project, with expected investments of up to $5.1 billion (Reuters 2023b). It is worth noting, however, that China's emerging international hydrogen policy involves both bilateral (e.g., with Australia, Germany, the

UK, and the US) and multilateral (with the IEA and IRENA) technological cooperation, as well as engagements with the EU, pioneering countries such as Germany, France, and Japan, and the Asia-Pacific Economic Cooperation, with a view to formulating standards for hydrogen-related technologies such as fuel cell vehicles (Quitzow and Gong 2023).

In the wake of growing competition with the US and Europe, China also continued to mobilize its economic might at home and abroad to maintain its leadership in the upstream segment of low-carbon supply chains. At the end of 2021, as anticipated, China established the China Rare Earth Group—merging three existing state entities—to consolidate its REEs industry and increase state control over the sector. In 2022, China expanded investments in lithium production and processing to increase self-sufficiency and keep pace with a growing demand for EVs batteries (Meidan, Hove, and Andrews-Speed 2023). The Ministry of Finance also increased the budget for stockpiling and improving the national reserve system of critical materials (ibid.). In addition, Chinese companies increased their investments in copper, lithium, and cobalt mines abroad, while the government strengthened relations with traditional partners. In 2022, Beijing upgraded the China–Indonesia Comprehensive Strategic Partnership, and Chinese companies—both mining companies and EV battery manufacturers, such as CATL—increased investments in the Indonesian nickel business (since 2013 China had already been the largest investor in Indonesia's mineral sector; Tritto 2023). Still in 2022, Beijing reached a new agreement with Eritrea, which had joined the BRI the previous year, for developing its mineral resources—especially copper, zinc, and gold—through a joint venture between the Chinese SOE Sichuan Road and Bridge Group and the Eritrean National Mining Corporation. In May 2023, then, Xi Jinping received the Eritrean president in Beijing for upgrading political and diplomatic relations with this country, which has a strategic position in the Horn of Africa. In May 2023, Xi Jinping also received in Beijing the president of the DRC to elevate their bilateral relations to a Comprehensive Strategic Cooperative Partnership, the highest level in China's diplomatic hierarchy. This move was also intended to review the $6 billion China–DRC 'infrastructure-for-minerals' agreement—signed in 2008 by the former DRC president—which was under scrutiny in the country, to plan new Chinese investments, and to increase DRC companies' stake in the existing copper and cobalt joint ventures with Chinese companies (Rolley 2023). Moreover, in 2022–2023, Chinese companies signed new deals with Latin America resource-rich countries, targeting the so-called 'lithium triangle', that is, the region of the Andes encompassed by the borders of Argentina, Bolivia, and Chile (Table 3.8).

Table 3.8 Chinese companies' new deals in the 'lithium triangle' (2022–2023)

Country	Date	Notes
Argentina	July 2022	China's Ganfeng Lithium agreed to buy the Argentina mining group Lithea for up to $962 million to secure access to more lithium resources for battery production
	April 2022	Chinese mining firm Zijin Mining Group decided to invest $380 million to construct a lithium industrial plant in Argentina
Bolivia	January 2023	The Bolivian government selected China's battery producer CATL and the cobalt miner China Molybdenum to develop its lithium reserves in a consortium with its state-owned companies
Chile	April 2023	China's EVs manufacture BYD agreed to build a lithium factory in Northern Chile(since 2018 the Chinese company Tianqi Lithium has 23% share in one of Chile's largest lithium producers, Sociedad Química y Minera)

Sources: Author's compilation from press sources (*Reuters* and *Financial Times*)

Unlike several previous deals, which were focused on the export of raw materials, the above-mentioned deals often included an industrial component—responding to the requests of the resource rich countries—and resulted in an important role for Chinese EV and battery companies (Table 3.8). The development of battery and EV supply chains has also become a pillar of Chinese engagement in the EU (Kratz et al. 2023). In 2022, for example, the Chinese CATL invested $7.6 billion to build a gigafactory in Hungary (previously, similar investments of approximately $6 billion had been made in France and Germany). This despite the fact that Beijing's overall investments in the bloc have continued to decline in recent years, also due to the tightening of the screening measures implemented in the EU, which have negatively impacted Chinese acquisitions of strategic assets, from semiconductor companies to critical infrastructure (Kratz et al. 2023). Chinese battery companies such as CATL are planning investments and partnerships with automakers in the US as well, although these ventures will be affected by the evolving status of the bilateral relations between Washington and Beijing (see also Chapter 6).

China's efforts to expand its grip on key low-carbon supply chains have developed in parallel with a sharpening of the country's armamentarium for green economic statecraft. This process also has to be placed in the context of the growing competition with Europe and rivalry with the US. At the beginning of 2023—only a few days after Japan and the Netherlands agreed to join the US chip restrictions on China—the Ministry of Commerce and the Ministry of Science and Technology issued new guidelines to (potentially) ban the export of several core PV technologies, such as solar wafers, in which China is the global leader. Then, in July 2023, the Ministry of Commerce issued new regulations to increase control over the export of gallium and germanium for reasons of national security. These two minerals—both on the EU and US lists of critical raw materials—are used for computer chips and for low-carbon technology such as solar PV and batteries (China supplies around 70% of the gallium consumed in the EU; Grohol and Veeh 2023). Finally, in late October 2023—a few days after the US increased controls on the export of advanced chips to Beijing and the European Commission announced its intention to start an anti-subsidy investigation into Chinese battery electric vehicles for passengers—the Ministry of Commerce announced that from the next December it would require export permits for certain graphite products with a view to safeguarding the country's 'national security and interests' (Euractive 2023). China is the world's largest producer and exporter of graphite, and this mineral too is on both the EU and US lists of critical raw materials, as it is a key component in the production of EV batteries.

4
The European Union
The Would-Be Green Leader

From energy markets to climate leadership and technological sovereignty

Energy played a central role in European integration following the Second World War. The initial steps included the creation of the European Coal and Steel Community (1951) and the European Atomic Energy Community (1957), which addressed both the primary energy source of that time (coal) and the most promising one (nuclear power). However, in the 1960s, the rapid diffusion of oil into the continent's economic system decoupled energy issues from the integration process. Only in the late 1980s did energy become intertwined again with this process. In 1988, the European Commission initiated the Internal Energy Market project under the Single European Act, aiming to establish a unified market. This period marked a significant shift in the standing of Western Europe within the international energy landscape. Oil prices dropped and the market and political power of the OPEC cartel was no longer a concern for consumers. In Europe, the Soviet Union, the UK, Norway, and the Netherlands ensured significant oil and gas supplies, while major countries like France and Germany pursued nuclear programmes. Renewables, except for hydropower in some European countries, remained a small niche. In a similar context, characterized by the diffusion of neoliberal thought, the European Commission embraced the idea of promoting closer integration among the energy markets of the member states by liberalizing the electricity and gas sectors.

In the 1990s, despite opposition from several member states and companies, the internal energy market project began to take shape. The initial directives for electricity and gas markets (collectively known as the First Energy Package) were enacted in 1996 and 1998, respectively. This period witnessed the consolidation of the market paradigm in the energy realm and the further relaxation of European relations with fossil fuel producers following the end of the Cold War. This was also the period in which the Maastricht

Green Superpowers. Andrea Prontera, Oxford University Press. © Andrea Prontera (2024).
DOI: 10.1093/9780191987304.003.0004

Treaty came into effect (1993), establishing the EU. During the negotiations for this treaty, the inclusion of an 'energy chapter' to extend EU-level competences beyond a predominant market focus was proposed, but this suggestion was not accepted by the member states (Schubert, Pollak, and Kreutler 2016). EU-level action in this field progressed in line with the regulatory state perspective, emphasizing the EU as a regulatory institution primarily focused on regulatory and market-based instruments. This approach informed the EU's internal and external energy polices in the ensuing decades, with the latter aiming to project EU rules and market perspective abroad through a combination of soft and regulatory power resources.

During the 1990s, the primary driving force behind both the internal and external green energy policies of the EU was climate change (e.g., Jordan et al. 2010; Wurzel and Connelly 2010; Rayner and Jordan 2016; Dupont, Oberthür, and Biedenkopf 2018). Building on its environmental competences expanded in the 1980s, the EU increasingly prioritized climate change in the aftermath of the 1992 Rio Summit. The promotion of renewables, reframed as a means of addressing global warming, gained momentum. The EU's approach to renewable energy policy largely involved soft coordination of member states' R&D policies (Hildingsson, Stripple, and Jordan 2012). More substantial initiatives were undertaken at the national level, with several EU countries adopting measures such as feed-in tariffs (e.g., Denmark, Germany, Portugal, Spain, Greece, Italy) or tendering schemes and auctions (e.g., UK, Ireland, France) to promote the deployment of renewables in the electricity sector. At the same time, the EU aimed to assume a leadership role in global climate policy (Oberthür and Roche Kelly 2008; Oberthür and Pallemaerts 2010; Wurzel and Connelly 2010). During negotiations on the UNFCCC and the 1997 Kyoto Protocol, the EU advocated for binding obligations for industrialized countries and pushed for more ambitious GHG emission reduction targets, positioning itself as a leader among industrialized economies. This early display of 'leadership by example' (Oberthür and Roche Kelly 2008, 36) marked the initial stages of the EU's green foreign energy policy, centred on the climate issue area, multilateral engagement, and employing a mix of diplomacy and soft power. This approach gained momentum in the 2000s, facilitated by the US withdrawal from the Kyoto Protocol, which provided an opening for EU global leadership, and the reinforcement of the EU's internal green energy policy. This latter trend overlapped with the construction of the internal energy market (the Second and Third Energy Packages were approved in 2003 and 2009, respectively), giving rise to a gradual 'greening' of the EU regulatory state. This involved the introduction of regulations, targets, and market-based instruments for climate and renewables within the EU regulatory and policy framework. This process began in the early 2000s when, in order to implement Kyoto targets, the European Commission issued the European Climate Change Programme, which resulted in the first EU legislative act for the promotion of renewables in the electricity sector (Directive 2001/77/EC, or RES directive) and in the launch of the Emission Trading System (ETS; Directive 2003/87/EC, or ETS Directive). The ETS, a market-based instrument, emerged as a cornerstone of EU climate policy, establishing a carbon price for large industrial and power plants and extending to domestic aviation in 2008.

In the mid-2000s, renewable energy gained increasing prominence on the EU agenda due to a confluence of internal and international developments (Solorio and Bocquillon 2017). After the 2004 Eastern enlargement, the EU's reliance on fossil fuel imports further increased. This trend was projected to escalate owing to declining EU fossil fuel production. In 2006, then, the gas crisis between Ukraine and Russia heightened concerns about security of supply, exacerbated by Putin's Russia adopting a more assertive foreign policy in the post-Soviet space. In the 2007 Communication An Energy Policy for Europe, the Commission stressed the role of renewables in climate mitigation and security of supply (European Commission 2007). It urged the EU to speak with 'one voice' in external energy relations. Although member states did not renounce their national foreign energy policies, especially in the fossil fuel sector, the Commission was able to develop bilateral energy and climate dialogues with third states. However, during the 2000s, its approach mainly centred on the externalization of the EU's internal rules and market approach. These efforts have been widely analysed by the scholarship in the context of EU 'external governance' (e.g., Herranz-Surrallés 2015; Weber 2017) and the 'external face' of the EU regulatory state (e.g., Goldthau and Sitter 2014, 2015a; Andersen, Goldthau, and Sitter 2016). This research explores how the EU combined regulatory power, capacity building, and (soft) conditionality, including within the framework of the European Neighbourhood Policy (ENP), to establish a Pan-European energy regulatory space spanning from Eastern Europe to the Southern Mediterranean.

At the end of 2008, during the first Barroso Commission (2004–2010), amid international climate negotiations focused on a post-Kyoto arrangement, the EU achieved consensus on the 2020 Climate and Energy Package. This package strengthened the integration between climate and energy policy and provided for a set of legislative measures that upgraded the RES and ETS directives and established the main EU regulatory and policy framework towards 2020. After its approval, the installed capacity of solar PV and wind energy in the EU experienced a significant surge: more than doubling from

2008 to 2012 and continuing to expand in the subsequent years (Figure 4.1). This growth was especially driven by feed-in tariffs and various supporting measures implemented in the larger EU member states. As a result, renewables gradually increased their role in the EU's electricity mix, with wind and solar energy rising from approximately 5% in 2010 to more than 20% a decade later (Table 4.1).

Moreover, in 2009, the Lisbon Treaty came into effect. It finally included a specific chapter on energy that explicitly mentioned, among the EU goals,

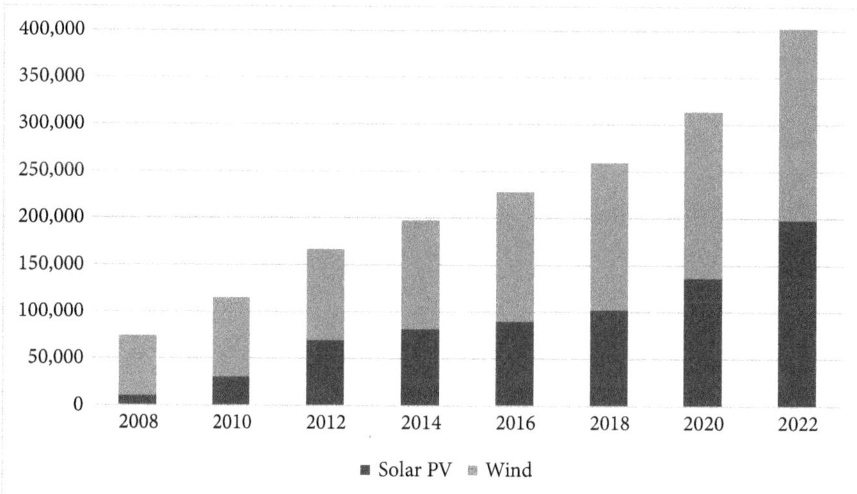

Figure 4.1 Solar and wind energy installed capacity in the EU (in MW, selected years)
Source: Author's elaboration from IRENA (2018, 2023a)

Table 4.1 Share of electricity production (in percentage) by source in the EU (2000–2022)

	2000	2005	2010	2015	2020	2022
Coal	30	28	23.7	24	12.7	15.9
Gas	12.6	17.8	20	13.8	20	19.9
Oil	8.7	6.9	5	4.4	3.6	3.5
Nuclear	32.7	31.7	29	27.3	24.8	21.9
Hydropower	13.3	10.6	12.6	11.6	12.5	10.1
Wind	1	2.3	4.7	9.1	14.4	15
Solar	0.1	0.05	0.7	3.5	5.2	7.2
Bioenergy	1.1	2.1	3.7	5.2	5.9	5.9

Source: Author's elaboration from Ritchie Roser and Rosado (2023)

the development of renewable energy sources, along with security of supply, energy efficiency, market integration, and energy networks. This chapter did not change the nature of EU competences in energy and climate matters, which are shared with member states, nor did it diminish the role of national governments in determining their energy mix and foreign energy policy. In other words, the EU remained a '*sui generis* actor in energy policy' (Youngs 2013, 421), with common rules that coexist with independent member states' policies. However, the Lisbon Treaty enhanced the role of the EU institutions in energy policymaking and improved the institutional architecture of EU foreign policy (Schubert, Pollak, and Kreutler 2016). In 2010, a Commissioner for Climate Action, along with the Directorate-General for Climate Action (DG CLIMA), was created. Subsequently, in 2011, the European External Action Service (EEAS) was established, helping the EU (gradually) strengthen its diplomatic capabilities, including in the area of energy and climate diplomacy (Herranz-Surrallés 2016; Dupont, Oberthür, and Biedenkopf 2018). This period coincided with the aftermath of the quasi-failure of the 2009 COP 15 at Copenhagen (e.g., Wurzel, Connelly, and Liefferink 2016; Dupont, Oberthür, and Biedenkopf 2018). The inability of the EU to influence the negotiation process at Copenhagen resulted from an evolving international landscape in which emerging powers were playing a more important role and the US–China relationship was decisive for the results of multilateral climate efforts. The EU, in this context, became a 'medium-size climate power', ranking behind Washington and Beijing (Oberthür and Dupont 2021, 1103). After Copenhagen, hence, the EU recalibrated its approach, which shifted from a strong focus on the multilateral UNFCC process towards raising climate change in high-level political forums such as the G7 and G20 and improved bilateral outreach to third countries. The EU assumed the role of a 'leadiator', characterized by more moderate policy objectives but greater emphasis on coalition- and bridge-building (Bäckstrand and Elgström 2013). This new role was instrumental in favouring the agreement reached in Paris in 2015. This shift was also accompanied by a stronger focus on capacity building and climate finance as key elements of the EU green foreign energy policy toolbox towards developing countries in the Global South (e.g., Dupont, Oberthür, and Biedenkopf 2018).

Under the second Barroso Commission (2010–2014), a new Communication on the EU external energy policy (The EU Energy Policy: Engaging with Partners Beyond Our Borders) and the revision of the ENP (A New Response to a Changing Neighbourhood: A Review of European Neighbourhood Policy) were issued in 2011. Although both policy documents were still imbued with the EU market approach and placed a strong emphasis on security of supply (in 2009, there was another gas dispute between Ukraine and Russia), the promotion of renewables abroad occupied a more prominent position with respect to the previous period. This resulted from the repercussions of the greening of the EU regulatory state, which showed its external reach. Moreover, by linking renewables to the notion of sustainability and sustainable development, the Commission recognized that these sources should be supported abroad by deploying all the EU foreign policy tools: bilateral and multilateral diplomatic engagements, the ENP framework for the neighbourhood, development aid and trade policy (where the EU has exclusive competences).

Sustainable development, promoted by several environmental NGOs and actors from the development community, has been a key principle of the EU development policy since the 1990s (e.g., Beringer 2019). With the Treaty of Amsterdam, which came into effect in 1999, sustainable development was institutionalized as an overarching objective of the EU action. The EU—which, along with its member states, is the world's largest donor—has been among the main promoters of the incorporation of this principle into the UN development framework until the Sustainable Development Goals were formulated as part of the post-2015 global development agenda. However, since the approval of the Agenda for Change in 2011—the new EU development framework that reflected the evolving international landscape characterized by the growing influence of emerging powers—the EU has increased its aid assistance for promoting climate goals and renewable energy abroad. From 2013 to 2021, the EU and its member states, which were already the biggest contributors of public climate finance to developing countries, more than doubled their disbursements, from around €10 billion to €23 billion (European Commission 2022a). In addition, the EU expanded the use of new forms of aid delivery, such as budget support and financial instruments like guarantees and blending. The European Investment Bank (EIB) also played an important role in this process. It increased its lending for climate-related investment in developing countries, a trend that will be further accelerated after the Paris Agreement. From 2012 to 2019, the EIB supported over €20 billion in climate action in these countries, and the proportion of EIB financing for climate action rose from 25% in 2015 to 31% in 2019 (EIB 2020).

With regard to trade policy, following the 'regulation through trade' logic, sustainability and climate change clauses were included in a new generation of EU (bilateral and inter-regional) trade deals. This move, favoured by the difficulties encountered in promoting environmental measures under the

WTO, contributed to further greening the external reach of the EU regulatory state. It is worth noting that, in this period, the EU was also involved in trade disputes over renewable energy and critical raw materials. The most important were those against China. The first (2012–2013) concerned an anti-dumping and anti-subsides investigation on Chinese PV modules and components, which was resolved bilaterally through negotiations between the Commission and Beijing (Lewis 2014). In the second case (2012–2014), the EU joined Japan and the US, challenging under the WTO the Chinese restriction on exports of REEs (Kalantzakos 2017a; see also Chapter 5). This latter dispute prompted the EU to upgrade its approach in the area of critical raw materials—a first initiative on that had been launched in 2008—improving cooperation with industry, member states, and potential foreign partners on this matter (European Commission 2013). However, significant steps in this direction were taken only a decade later (as detailed below).

Finally, an additional crucial event that contributed to the transformation of the EU green foreign energy policy in this period was the 2008 global financial crisis. The crisis affected state–market relations within Europe, prompting the emergence of a more interventionist approach by the EU beyond its traditional regulatory state focus. In particular, the crisis marked the beginning of a new era, characterized by a (green) industrial policy revival across the continent (e.g., Tagliapietra and Veugelers 2023). The EU gradually abandoned its previous approach based on liberalization and market-oriented (horizontal) policy measures—centred on setting the right framework conditions for market actors and stimulating R&D—to embrace a more targeted and direct approach to sustaining strategic sectors. This nascent EU-level industrial policy was further strengthened in the following years as a result of a new consensus between France and Germany (Di Carlo and Schmitz 2023). This change was particularly evident in the field of climate and energy, where the nucleus of an EU catalytic state began to emerge, focusing on energy networks and low-carbon technologies (Prontera and Quitzow 2022). These technologies were then more explicitly viewed not only as essential ingredients for mitigating climate change and reducing EU dependence on fossil fuels but also as a means to reinvigorate European economic development in the face of an increasingly competitive international environment. In turn, this shift in state–market relations affected the EU's green foreign energy policy and green power strategy. The EU embraced a more interventionist stance in its external engagements, increasing the use of economic tools in line with the catalytic power perspective.

The reframing of renewables, the sharpening of the EU green regulatory state, the shift in state–market relations towards more interventionist approaches with the consolidation of the EU catalytic state, and the expanded use of economic tools for external action in line with the catalytic power perspective all continued under the Juncker Commission (2014–2019). Moreover, in this period, for the first time, the notion of energy transition entered the EU agenda. In 2015, after the Russian annexation of Crimea and the war in Eastern Ukraine, the Commission issued the Energy Union Strategy (A Framework Strategy for a Resilient Energy Union with a Forward-Looking Climate Change Policy). Although, initially, this strategy was centred on security of gas supply—in 2015, Russia accounted for about 40% of EU gas imports—it was soon broadened to climate policy, renewables, and low-carbon technology. Unlike previous strategies, such as the 2007 An Energy Policy for Europe, this time the EU explicitly referred to the notion of 'energy transition'. This transition was envisioned to be facilitated by promoting a more appropriate market design and regulatory framework, alongside fostering investments in low-carbon technology and energy infrastructures within the EU and its neighbouring regions. The Commission urged the EU to improve its 'technological leadership' and become 'the world leader in renewable energy' (European Commission 2015, 15). However, it also emphasized the need for the energy transition to be 'just and fair', and this required appropriate social measures (European Commission 2015, 17).

Although the EU was on track to meet its 2020 objectives, promoting more ambitious green action at home was more challenging in light of the economic crisis—which prompted several EU countries to reduce supporting measures for renewables (Gürtler, Postpischil, and Quitzow 2019; Prontera 2021)—and increasing internal opposition to stringent climate policy, especially from Central and Eastern EU member states (e.g., Skovgaard 2014). In 2014, the EU agreed on the 2030 Climate and Energy Framework to upgrade the 2020 package with new targets for 2030. This agreement also provided for dedicating at least 20% of the EU 2014–2020 multi-annual budget (including the part related to external action) to climate-related actions. The concrete measures for implementing the 2030 Climate and Energy Framework, however, were approved only in 2018. Also, in 2019, the Commission launched the Clean Energy for All Europeans Package, which aimed at reforming the electricity market design to better enable the energy transition.

Overall, the innovations enacted under the Juncker Commission further expanded the scope of the EU green foreign energy policy beyond the original focus on climate diplomacy, soft power, and regulatory tools. Yet the different elements of this policy were still poorly connected at the end of the 2010s. It was only with the appointment of the von der Leyen Commission in 2019 that important efforts were made to upgrade the EU's green internal and external action and increase its coherence. With the launch of the European Green Deal, in particular, the von der Leyen Commission placed the energy transition at the centre of the EU's internal and international agenda for the 2019–2024 period. As in the past, climate issues were chosen to relaunch the EU integration after a difficult period characterized, among other things, by Brexit, the rise of neo-populist and Eurosceptic political parties in both Western and Eastern Europe, and the growing tensions in transatlantic relations under the Trump administration. This agenda was also supported by the increasing political mobilization for climate action, spearheaded by the Friday's for Future movement, and backed by important political parties in the European Parliament, such as the Party of European Socialists and the Greens (e.g., Parks, Della Porta, and Portos 2023). The European Green Deal (EGD) merged the EU's industrial, green, and climate policies. It was presented as a 'new growth strategy' with the ambitious goal to make Europe the first climate-neutral continent by 2050 (European Commission 2019, 2). However, in line with the Energy Union Strategy, it also stressed the need to promote a just transition—i.e., a fair and inclusive process 'to leave no one behind'—providing for a Just Transition Mechanism and a Just Transition Fund (European Commission 2019, 16). Also, it reasserted the necessity of continuing reforming the EU regulatory and market framework to foster the energy transition and to increase financial and industrial support to low-carbon technology, providing for a Green Deal Investment Plan. Moreover, the EGD provided for the establishment of a Carbon Border Adjustment Mechanism (CBAM), a unilateral measure to leverage the EU market to promote decarbonization abroad while avoiding 'carbon leakage' and defending the EU industry.

Shortly thereafter, the Covid-19 began to impact Europe. Contrary to initial concerns, the crisis did not reverse the new EU's green agenda. Building on the Green Deal Investment Plan, the EU introduced the NextGeneration EU stimulus package in 2020, with the Recovery and Resilience Facility (RRF) at its core. This novel, albeit temporary, financial instrument enabled the European Commission to raise funds on financial markets on behalf of the EU—confronting the historical taboo about EU-level borrowing—and make them available to sustain member states' responses to the pandemic through the implementation of National Recovery and Resilience Plans (Bocquillon, Brooks, and Maltby 2023). With its unprecedented size (around €724 billion, of which €385.5 billion in loans and €338 billion in grants), the RRF significantly expanded the EU's spending capacity. Moreover, 37% of RRF funds were earmarked for climate-related investments and reforms in the member states, which was higher than the equivalent figure set for the

EU 2021–2027 multi-annual budget, which was 30% (10% more than the 2014–2020 period), and significantly more than the €22 billion of the EU post-2008 financial crisis green stimulus (Quitzow et al. 2023). In 2020, with the Taxonomy Regulation (Regulation 2020/852/EU) and the launch of the EIB Climate Bank Roadmap, the EU also increased its efforts to greening the European financial system. It is worth noting that the Taxonomy Regulation allowed for supporting natural gas and nuclear energy on the request of some member states, such as France, where nuclear power covers more than 60% of electricity production. The EIB, however, with its new Energy Lending Policy, issued in 2019, committed to stop fossil fuel investments from 2021 and to increase its level of support for climate action and environmental sustainability to exceed 50% of its overall lending activity (inside and outside the EU) by 2025 (EIB 2020).

In 2020, to implement the EGD, the EU introduced several key strategies and plans. The first was the Hydrogen Strategy, which primarily focused on green hydrogen, aiming to shift away from hydrogen production based on fossil fuels. Simultaneously, the EU launched the Offshore Renewable Energy Strategy and an Action Plan on Critical Raw Materials. The latter plan represented an update and improvement of the previous EU strategy on raw materials. It specifically focused on minerals such as lithium, cobalt, nickel, and REEs, which are fundamental to the EU's green industrial turn. The plan also targeted actions to enhance the diversification and resilience of supply chains, a concern magnified by the Covid-19 crisis. The pandemic exposed the vulnerability of the EU to shocks and shortages in critical supply chains.

On January 2021, then, the European Council fully embraced the external dimension of the EGD, stressing that the EU's energy diplomacy would aim, as its primary goal, to accelerate the global energy transition, promoting energy efficiency and renewable technologies (European Council 2021). The same year, the EGD commitments on climate neutrality by 2050 became binding with the European Climate Law (Regulation 2021/1119), which also upgraded the previous 2030 goals, setting the intermediate target of reducing EU GHG emissions by at least 55% compared to the 1990 levels. To implement these new climate targets for 2030, the Commission presented a set of legislative proposals, the so-called Fit for 55 Package. This move was instrumental in reasserting the EU's leadership in the lead up to 2021 COP 26 in Glasgow. The Fit for 55 Package also included concrete proposals for the CBAM mechanism, which entered into force in 2023.

In 2021, the EGD permeated the review of the EU trade policy (An Open, Sustainable and Assertive Trade Policy) and the new EU industrial strategy (Building a Stronger Single Market for Europe's Recovery). Besides, both these initiatives reflected the awareness of a more challenging international environment that called the EU to deploy its capabilities in order to promote its 'technological sovereignty' in the green and digital sector, defend the single market, tackle dependencies in critical supply chains, and foster the bloc's 'open strategic autonomy' (European Commission 2020, 2021). The notion of open strategic autonomy, in particular, had already emerged under the Juncker Commission as a way to find a balance between the EU's preferences for an open global economy with the challenges posed by a more hostile international environment characterized by transatlantic tensions (especially during the Trump era), a growing rivalry between the US and China, and more assertive foreign policies from Beijing and Moscow (Damen 2022). The perspectives of technological sovereignty and open strategic autonomy imply an additional shift in state–market relations towards a more interventionist role for the EU. This more interventionist stance at home was paralleled by efforts to improve the geopolitical posture of the EU abroad, especially thanks to a more focused mobilization of the bloc's economic tools. This manifested in the launch, on December 2021, of the Global Gateway Strategy backed by the Neighbourhood, Development and International Cooperation Instrument-Global Europe (NDICI-Global Europe). With an overall allocation of €80 billion, of which 30% was earmarked for climate goals, the NDICI-Global Europe centralized under a common framework all the main financial tools for the EU's external action for the period 2021–2027. The objective was to better promote the bloc's foreign policy priorities. The Global Gateway Strategy set an ambitious goal of mobilizing around €300 billion of investment by 2027. It aligned with initiatives launched in June 2021 by the G7 and US, such as the Build Back Better World initiative aimed at countering China's BRI in the Global South. Additionally, it echoed the commitments made by the EU at COP 26, focusing on addressing the climate crisis through the development of clean and resilient infrastructures with high environmental, social, and governance standards (European Commission-High Representative of the Union for Foreign Affairs and Security Policy 2021).

The decisive push to upgrade the perspectives of technological sovereignty and open strategic autonomy and strengthen the geopolitical turn of the EU emerged prominently in the aftermath of the 2022 Russian invasion of Ukraine. At the beginning of her mandate, Ursula von der Leyen had already embraced the narrative of a 'geopolitical Commission', but this shift took on clearer contours following the events of 2022 (Siddi and Pandrin 2023; Herranz-Surralles 2024; see also European Commission 2022b). In the wake of the war, the EU, with the issuance of the REPowerEU Plan in May 2022, made a historic decision to phase out its dependence on Russian fossil fuels 'as soon as possible' (European Commission 2022c, 1). This paradigm shift became conceivable only after the overt Russian invasion of Ukraine, given that in 2021, about 43% of the EU's gas imports and 27% of its oil imports came from Russia (a similar decision was not even considered after the 2014 Moscow annexation of Crimea). With the REPowerEU Plan, the EU raised its ambitions in the area of renewables—upgrading the previous targets for 2030—and planned a series of measures to speed up the deployment of wind and solar energy, as well as green hydrogen. After the invasion, renewables were framed even more prominently than before as a means to increase energy security and achieve decarbonization objectives. The entire EU energy transition became 'more explicitly securitised' (Goldthau and Youngs 2023, 5), opening the way for additional interventionist measures both domestically and abroad.

Thanks to the combination of EU-level and national-level actions (e.g., Prontera 2024), EU dependence on Russian gas was quickly reduced: one year after the beginning of the war, EU imports from Moscow dropped to around 15%, with increasing supplies coming from the US, Qatar, Norway, Algeria, and Azerbaijan (Eurostat 2023). On the other hand, as a result of EU and international sanctions, imports of oil products from Russia fell to 6% in the same period, while coal imports were completely halted (Eurostat 2023). The EU also reduced its imports of nickel from Russia, although they were not affected by a specific ban, increasing supplies from the US, Australia, and Norway (ibid.).

Moreover, although the Russian invasion spurred a revival of EU (and member states) initiatives in the area of fossil fuels, the EU's energy transition was further pushed by the war. This, however, was accompanied by increasing worries in Brussels and national capitals to switch from overdependence on Russian fossil fuels to overdependence on China for delivering the energy transition. Indeed, Beijing is the main EU's supplier for several critical minerals and low-carbon technologies. In the area of REEs and PV equipment, the EU's dependency on China is close to 90% and 80%, respectively (European Commission 2023a; Grohol and Veeh 2023). In the EVs and battery sector as well, the EU is highly dependent on China (for more than 25%), an issue that was even more pressing after the bloc agreed, in 2022, to ban fossil fuel–based combustion engines by 2035. However, in general, the EU recognized its heavy reliance on imports from a few countries in the green realm. Hence, in 2023, it launched a series of initiatives to upscale its industrial capacity in strategic sectors, reduce its vulnerability in low-carbon supply chains (from PV modules, wind turbines, and batteries to critical minerals), and improve its green international posture. These include the

Green Deal Industrial Plan, the Net Zero Industry Act (NZIA), which was also a response to the US Inflation Reduction Act (IRA), and the Critical Raw Materials Act (CRMA; a European Chips Act for the semiconductor sector was also approved in 2023). At the same time, as highlighted in its 2023 European Economic Security Strategy, the EU preferred to frame its vulnerability in the area of the energy transition as a problem of 'de-risking' and continued to reaffirm its willingness to reform and strengthen the WTO (European Commission-High Representative of the Union for Foreign Affairs and Security Policy 2023). To be sure, relations with Beijing became tenser after the beginning of the war. After several years of negotiations, in December 2020, the EU and China reached a political agreement on the EU–China Comprehensive Agreement on Investment (CAI). This deal, however, which was already frozen by the European Parliament in 2021, was gradually abandoned in 2022. Moreover, on October 2023, the European Commission decided to start an anti-subsidy investigation into the imports of battery electric vehicles from China.

In the following sections of this chapter, I first illustrate in more detail the greening of the EU regulatory state, emphasizing particularly its external repercussions until (circa) 2020. Then, I focus on the innovations introduced since the 2008 economic crisis, which drove the rise of the EU catalytic state and the emergence of a catalytic green power strategy. Finally, I illustrate the post-2020–2021 developments, characterized by the external ramification of the EGD's implementation and by the innovations introduced after the war. Although the EU has continued to pursue its leadership in the field of multilateral climate diplomacy, these developments point to a further expansion of the bloc's green foreign energy policy towards the industry, trade, and investment issue areas, a closer integration of its different tools in these fields, a stronger emphasis on minilateral and bilateral modes of engagement, and to an upgrading in the use of its economic might, which falls short of embracing a green economic power strategy.

Greening the EU regulatory state

The greening of the EU regulatory state has been instrumental in expanding the EU's green foreign energy policy beyond its original focus on climate leadership and soft power. This process was initiated in the early 2000s and continued for the following two decades, during which the construction of the single energy market overlapped (first) and merged (later) with the development of the EU climate and renewable energy policy framework.

Figure 4.2 (Section A) briefly summarizes the main steps and components of this path, with a focus on emission reduction and renewable energy. The targets and the achievements in these areas are further specified in Table 4.2.

As mentioned, the starting point in this process was the RES directive and the ETS directive. Later, under the 2020 Climate and Energy Package, the Renewable Energy Directive (Directive 2009/28/EC, or RED I directive) established a common framework for promoting renewables not only in the electricity sector but also in the transport, heating, and cooling sectors, setting an overall EU target of 20% by 2020. This package included a revision of the ETS and a decision on sharing the efforts among member states to reduce GHG emission not covered by this market mechanism. As requested by the Commission, the RED I directive also provided for mandatory targets for member states in terms of overall share of renewables in gross final energy consumption. However, not all the Commission's proposals for the 2020 Climate and Energy Package were endorsed by member states. In particular, the Commission failed in its attempts to harmonize the national support schemes for renewables in the electricity sector according to the tradable green certificates model (Solorio and Bocquillon 2017). This proposal was opposed by important member states, such as Germany and Spain, which had very successfully expanded the deployment of renewables due to feed-in tariffs. In 2014, however, the EU enacted new guidelines on state aid for environmental protection and energy, which prescribed that feed-in tariffs were no longer regarded as permissible state aid. These supporting mechanisms had to be gradually replaced by tools granting more relevance to market signals, such as feed-in-premiums and tradable green certificate schemes.

The 2030 Climate and Energy Framework included a further revision of the ETS, an effort-sharing regulation, and a new directive on renewables, the RED II (Directive EU 2018/2001), which set a target of at least a 32% share for renewable energy at the EU level by 2030, but without nationally binding objectives. In 2018, however, the EU also enacted the so-called Governance Regulation (Regulation 2018/1999), which required member states to draft National Energy and Climate Plans synchronized with cycles under the Paris Agreement and that enhanced the influence of the European Commission in the formulation and implementation of member states' energy policies (e.g., Knodt, Ringel, and Müller 2020). At the same time, the measures introduced by the 2019 Clean Energy for All European Package intervened in the design of the EU electricity market, allowing more flexibility to accommodate an increasing share of renewable energy in the grid. The creation of a European market design for green energy continued with the launch, in 2020–2021, of the framework for offshore renewable energy and for hydrogen and decarbonized gas (finalized in 2023), and with the approval in 2022 of the new Guidelines on State Aid for Climate, Environmental Protection and Energy (CEEAG), which expanded the possibility for member states to support investments in all the low-carbon technology that can assist in delivering the EGD (Figure 4.2). The latter, together with the European Climate Law, the measures provided by the Fit for 55 Package—in particular the reform of the ETS, a new effort-sharing system, the RED III directive, and an additional reform of the electricity market design (all finalized in 2023)—further improved the greening of the EU regulatory state and upgraded the EU climate and renewable energy targets (Table 4.2). Following proposals already advanced in the 2022 REPowerEU Plan, in which the Commission suggested a 45% target, the RED III directive, in particular, increased the 2030 target for renewables from 32% to 42.5%, to be provided as a compulsory contribution by the member states, ensured by the existing Governance Regulation (with another 2.5% that can be achieved through further voluntary contributions or through pan-European measures; Table 4.2). To achieve this new target, the EU will have to install more than 100 GW of new wind and solar capacity every year. To facilitate this huge expansion, the RED III directive introduced new common rules to accelerate approval procedures for the installation of renewables. It also set more ambitious targets in the area of green hydrogen (in industry), as well as provisions to facilitate cross-border cooperation projects involving more member states (e.g., joint offshore wind projects).

However, to understand how this internal evolution has influenced the EU green foreign energy policy it is necessary to examine the 'external face' of the EU green regulatory state, which is composed of several elements. First, there are the regulatory and policy regimes the EU has built to export its internal rules and objectives in its neighbourhood. However, the 'external face' of the EU green regulatory state also includes the EU's bilateral and multilateral diplomacy and the use of trade deals to foster globally the bloc's preferred rules and principles.

By exercising (positive and negative) conditionality, the EU has used the attractiveness of its market and its regulatory power to export its rules to its surrounding regions (e.g., Schimmelfennig and Sedelmeier 2004; Lavenex 2014). By the late 1990s, the Commission was already promoting the Energy Charter Treaty to extend free trade rules and investment protection provisions to the fossil fuel sector of the post-Soviet countries (Prontera 2017). As anticipated, however, it was in the early 2000s—in parallel with the enhancement of the internal energy market—that the EU increased its efforts to create a common regulatory space with its neighbours by deploying a complex mix of minilateral and bilateral tools: from legally binding treaties, such as the

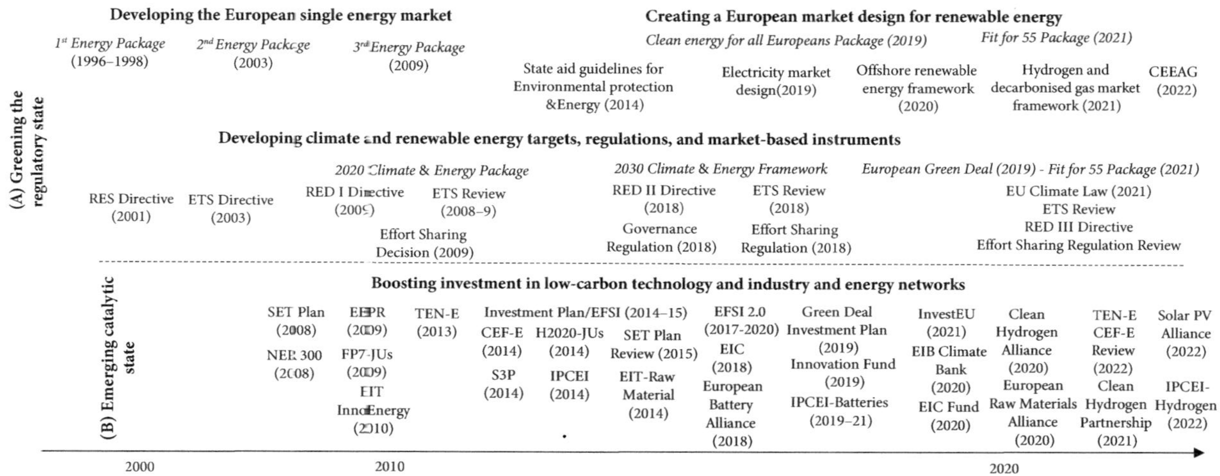

Figure 4.2 The greening of the EU regulatory state (A) and the rise of the catalytic state for low-carbon technology and energy networks (B)

Notes: (A) CEEAG (Guidelines on State aid for climate, environmental protection and energy); (B) SET Plan (Strategic Energy Technology Plan), and NER 300 = support for low-carbon innovation with financial assistance and multi-stakeholder platforms; EEPR (European Energy Programme for Recovery) = financial assistance to energy networks and offshore wind projects; FP7–JUs (7th Framework Programme-Joint Undertakings) = research funding and public–private partnerships in low-carbon technology fields (upgraded in 2014 with Horizon 2020 and, in 2021, with Horizon Europe); S3P (Smart Specialization Platforms) = public–private platforms to support regional authorities' innovation strategy; European Institute of Innovation and Technology (EIT) = the EIT, established in 2008, works in coordination with Horizon programmes supporting research, innovation, and promoting industrial partnerships; it focuses on different sectors, including renewable energy (InnoEnergy) and raw materials; TEN-E (Trans European Energy Networks for Energy) = governance framework for planning and supporting cross-national energy networks, the so-called Projects of Common Interest (PCIs, mainly in the electricity and gas sector); CEF-E (Connecting Europe Facility for Energy) = works in coordination with TEN-E offering financial support to PCIs; IPCEI (Important Project of Common European Interests) = tool to support research, development, and innovation for cross-national projects combining state aid and private capital; SET-Plan review = upgrading of the SET-Plan, better focus on renewable energy technology (e.g., PV, wind, batteries) in line with the Energy Union Strategy (2015); EIC (European Innovation Council)/EIC Fund = support for research and innovation for the green and digital transition; EFSI 2.0 = upgrading of the EFSI, better focus on green and climate projects; Innovation Fund = upgrading of NER 300 for supporting low-carbon technology; InvestEU = successor and upgrading of EFSI to support investment in innovation and industrial capacity through guarantees and cooperation with the EIB and national development banks (linked to the Green Deal Investment Plan); EIB Climate Bank = roadmap linked to the EU taxonomy to align EIB financing with the Paris Agreement and European Green Deal; Clean Hydrogen Partnership = public–private partnership, upgrading of the Fuel Cells and Hydrogen JUs; TEN-E Review = revision of the previous 2013 TEN-E scheme, focus on electricity networks, wind offshore corridors, and hydrogen corridors (including those involving non-EU countries). CEF-E Review = refocused on electricity and cross-border renewable energy projects

Sources: Adapted and updated from Prontera and Quitzow (2022)

Table 4.2 Main climate and renewable energy targets and achievements in the EU

	2020 Climate and Energy Package	Achieved in 2020 (*)	2030 Climate and Energy Framework	European Green Deal/ European Climate Law
Climate	20% reduction in GHG emissions from 1990 levels by 2020	24% (**)	40% reduction in GHG emissions from 1990 levels by 2030	55% reduction in GHG emissions from 1990 levels by 2030 Climate neutrality by 2050
Renewable energy	20% of renewables in the EU final energy consumption by 2020	22%	32% of renewables in the EU final energy consumption by 2030	Fit for 55 Package 2021: 40% of renewables in the EU final energy consumption by 2030 RED III directive 2023: 42.5% of renewables in the EU final energy consumption by 2030 (^)

Notes: (*) = data from EEA (2021); (**) = this data is from 2019, as the data for 2020 is affected by the consequences of the Covid-19 crisis; (^) = another 2.5% can be achieved through voluntary contributions by the member states or through pan-European measures (45% by 2030 was the overall target proposed by the Commission in the 2022 REPowerEU Plan)
Source: Author's elaboration

European Economic Area (EEA) and the Energy Community Treaty (EnC), and trade agreements to looser forms of political dialogue and cooperation. The EEA, established in 1994, provided for the export to third states of EU legislation in the area of the single market, energy, and climate. The EnC, which came into effect in 2006, was created to promote the adoption of the EU legislation in the countries involved in the EU enlargement policy and to widen the sphere of the internal energy market on the EU's south-eastern periphery. With the establishment of the ENP in 2004—aimed at the countries in the Southern Mediterranean and Eastern Europe—energy issues were also included in this general framework for the externalization of EU policies (Batzella 2018). At first glance, the ENP seemed to apply the logic of conditionality adopted for the enlargement countries. It envisaged regional forums for political dialogue and diplomatic exchanges, such as the Union for the Mediterranean and the Eastern Partnership, bilateral institutional structures to sustain and monitor cooperation between the EU and third countries (e.g., Association Councils, Action Plans, and Association Agendas), technical and financial assistance (capacity building), and bilateral trade agreements offering (selected) access to the EU market (e.g., Association Agreements and

Deep and Comprehensive Free Trade Areas). However, the ENP lacked the promise of EU membership; it was based only on a soft version of conditionality aimed at a progressive convergence or approximation towards the EU's rules, with varying levels of ambition dependent on the third countries' preferences (Lavenex 2011, 2014; Börzel and Risse 2012; Buzogány 2018).

Over the time, this complex institutional architecture—led primarily by the Commission's Directorate General for Neighbourhood and Enlargement Negotiations (DG NEAR), the Directorate General for Energy (DG ENER), and the Directorate General for Trade (DG TRADE)—has incorporated the EU's green turn in its scope and priorities (Table 4.3). The EU had established an energy dialogue with Norway, which was included in the EEA, as early as 2002. Moreover, Norway and Iceland joined the EU ETS in 2008. In addition, Norway has gradually become involved in several EU initiatives on renewables and electricity networks, such as the Baltic Energy Market Interconnection Plan (2009), the North Sea Countries' Offshore Grid Initiative (2010), and the North Seas Energy Cooperation (launched in 2016 and upgraded in 2020). These initiatives have increased regulatory cooperation and market integration between the EU and Norway, which is linked to the EU power grid. With regard to the Western Balkans, after the 2020 Climate and Energy Package, the promotion of renewables was included in the EnC. After 2015, the EnC also established a platform of national regulators to promote the implementation of EU measures for renewable energy. Moreover, the EU has launched accession negotiations with several candidate countries in the Western Balkans, which will need to adopt the EU climate and energy policy frameworks with a view to joining the EU. Overall, although problems with the implementation of the EU market and energy rules persist (e.g., Energy Community 2022), countries in the Western Balkans have gradually become embedded in a dense network of regulatory cooperation with the EU. This made it possible to continue the externalization of the EU green regulatory state (see below).

After the 2006 Russia–Ukraine gas dispute, the EU also sought to expand the EnC to include Turkey and those North African countries covered by the southern dimension of the ENP, that is, the Southern Neighbourhood. However, neither Ankara nor the other governments in this region—especially gas producers such as Algeria, Libya, and Egypt—have shown interest in a similar project that would have limited their energy sovereignty. Energy cooperation in the Southern Neighbourhood, including in the area of renewables, has therefore been channelled primarily through the Union for the Mediterranean, transgovernmental networks that were established to mirror the internal development of the EU energy market—i.e., associations of regulators (Medreg) and transmission system operators (Med-TSO) from the EU and Southern Mediterranean countries—and bilateral engagements in the context of the ENP (Table 4.3). These initiatives shared the same set of common goals: promoting regulatory approximation towards the EU rules and exporting the EU model of energy market governance. These goals were pursued both at the regional and bilateral levels, where the ENP action plans envisaged assistance for reforms in the energy sector focused on liberalization and the establishment of independent regulators. Regulatory approximation also aimed to promote energy infrastructures and interconnections between the two shores of the Mediterranean Sea.

Despite the very poor achievements in exporting rules and the plan for interconnectivity, the EU's approach changed only slightly in the aftermath of the Arab Spring (e.g., Cambini and Rubino 2014; Herranz-Surrallés 2018). At the beginning of the 2010s, however, the promotion of renewable energy began to feature more prominently on the agenda of EU engagement in the Southern Neighbourhood. Certain industrial projects for importing renewable energy into the EU (e.g., the Mediterranean Solar Plan and the Desertec) were also proposed in this period. These initiatives were driven by EU energy security considerations and were designed according to the bloc's market approach. Although they ultimately failed—due to changing market conditions and lack of support from EU member states and third countries—they represented the first effort by the EU to create corridors for importing renewable energy into the continent. Moreover, after the launch of the Energy Union Strategy and the Paris Agreement, the EU increased its efforts to promote the energy transition in the region. For example, a new platform for cooperation and dialogue on renewable energy was established under Union for the Mediterranean, and renewable energy was more firmly embedded in the ENP framework. In addition, the EU increased its focus on renewables in the bilateral engagements established in the forms of memorandums of understanding (MoUs) envisaging Strategic Energy Partnerships with Algeria and Egypt (Table 4.3).

In the Eastern Neighbourhood, the EU has used the Eastern Partnership (established in 2009 and upgraded in 2017) as a regional platform for dialogue on market integration, energy infrastructure, and regulatory convergence. Regulatory approximation in the region has also been pursued by the EU in the context of the bilateral engagements of the ENP framework. Moreover, bilateral energy relations have been further developed by the EU with Ukraine and Azerbaijan. All these initiatives were originally conceived with a focus on security of supply, particularly in the natural gas sector. Nevertheless, after the approval of Energy Union Strategy and the

Table 4.3 Exporting the green EU regulatory state in the neighbourhood: regional (minilateral) and bilateral initiatives (2000–2020)

	Regional	**Bilateral**
European Economic Area (EEA) (Iceland, Norway, Liechtenstein)	EEA (1994): legally binding treaty for exporting of EU legislation linked to the single market as well as energy, environment, and climate Norway and Iceland participate in the EU ETS since 2008	Energy Dialogue: Norway (2002) on energy market, infrastructures, fossil fuel, renewables, and energy efficiency
Western Balkans (Albania, Bosnia and Herzegovina, Kosovo, North Macedonia, Montenegro, Serbia)	Energy Community Treaty (2006): legally binding treaty for exporting the EU energy legislation (e.g., RES directive incorporated in 2012) Energy Community Renewable Energy Coordination Group (2015): platform for supporting the implementation of measures on renewable energy	Enlargement Policy—Accession negotiations with candidate countries: Montenegro (2012), Serbia (2013), Albania (2018), North Macedonia (2020) Potential candidate countries: Bosnia and Herzegovina, Kosovo
Southern Neighbourhood (Algeria, Egypt, Israel, Jordan, Lebanon, Libya*, Morocco, Palestine, Syria**, Tunisia)	Union for the Mediterranean (2009) Platform on Renewable Energy and Energy Efficiency (2015): dialogue and policy exchanges on renewables and energy efficiency measures Medreg (2007): forum of energy regulators (harmonization of the regulatory framework, market integration, energy trade, regulatory measures for the development of renewables) Med-TSO (2012): forum of Transmission System Operators (harmonization of the regulatory framework, market integration, electricity interconnections)	ENP Framework—Association Agreements with Tunisia (1998), Morocco (2000), Israel (2000), Jordan (2002), Egypt (2004), Algeria (2005), Lebanon (2006): provisions on regulatory approximation, cooperation on renewable energy and energy efficiency MoU/Strategic Energy Partnerships (cooperation and dialogue in the fossil fuel sector and renewable energy): Egypt (2008, 2018), Algeria (2013, 2015)

Continued

Table 4.3 *Continued*

	Regional	Bilateral
Eastern Neighbourhood (Armenia, Azerbaijan, Belarus ***, Georgia, Moldova, Ukraine)	Eastern Partnership (2009; upgraded in 2017): dialogue on convergence towards the EU market, market integration, infrastructures, renewable energy, and energy efficiency Eastern Partnership-EU4Energy (since 2016): capacity-building programme (improving the energy policy and regulatory framework, including in the area of renewables) Energy Community Treaty (see above): Moldova (2010), Ukraine (2011), Georgia (2017)	EU–Azerbaijan Partnership and Cooperation Agreement (1999): cooperation on energy trade, energy infrastructures, and renewable energy (negotiations for a new agreement since 2016, still pending) ENP Framework—Association Agreements/Deep and Comprehensive Free Trade Area with Georgia (2014), Moldova (2014), Ukraine (2017): regulatory approximation with the EU energy rules including in the area of renewables ENP Framework—Comprehensive and Enhanced Partnership Agreement with Armenia (2017): regulatory approximation with the EU energy rules including in the area of renewables MoU/Strategic Energy Partnerships with Ukraine (2005, 2016) and Azerbaijan (2006): cooperation on security of supply, infrastructure, and renewable energy

Notes: Turkey is involved in the EU enlargement policy since 2005, however, accession negotiations were suspended in 2018. (*) = Libya has never concluded an Association Agreement with the EU and it has only an observer status in the Union for the Mediterranean; (**) = Syria has suspended its membership to the Union for the Mediterranean in 2011; (***) = EU cooperation with Belarus was suspended in 2020

Sources: Author's elaboration

Paris Agreement, regulatory reforms (and capacity building) for renewable energy were more firmly embedded in the EU external engagements. However, problems remained regarding the EU's ability to promote its regulatory preferences in this region. For example, negotiations to renew the 1999 EU–Azerbaijan Partnership and Cooperation Agreement were launched in 2016, but are still pending. The most important success for the EU in the Eastern Neighbourhood has been the accession of Ukraine, Moldova, and Georgia to the EnC. This has made it possible to upgrade regulatory cooperation, harmonization, and market integration, especially with Ukraine and Moldova, which are linked to the EU power grid. Since the 2022 Russian invasion, the EnC has also proved useful in providing support to the Ukrainian energy sector. Moreover, in June 2022 the EU granted the status of candidate countries to Ukraine and Moldova, and of potential candidate to Georgia. A choice confirmed on December 2023, when then the European Council decided to open accession negotiations with Kyiv and Chișinău. This might possibly increase the EU's ability to export its rules to the Eastern Neighbourhood.

While it was supporting the externalization of the green regulatory state in its near abroad, the EU also increased its diplomatic efforts to promote its green approach globally. To be sure, the ENP framework and the one for the Western Balkans also function as diplomatic venues for promoting the EU's approach to the energy transition. However, since the mid-2000s, the Commission (especially through DG ENERGY and DG CLIMA) has launched an increasing number of bilateral dialogues and partnerships on climate and energy beyond the EU's neighbourhood; these initiatives also covered renewable energy, thus reflecting the integration between these issues in the EU (Table 4.4). This diplomatic cooperation also targeted emerging powers, particularly China and India. An EU Russia energy dialogue was established as early as 2001, but it was abandoned in 2014 in the wake of the Russian annexation of Crimea. In all these cases, the EU deployed diplomatic tools and sought to exercise soft power to promote abroad its view on energy transition and the appropriate policies to promote it. This strategy has been particularly evident for the EU ETS. The European Commission has supported the diffusion of this market mechanism abroad both in multilateral negotiations and through bilateral engagements (Dupont, Oberthür, and Biedenkopf 2018). At the bilateral level, the EU has engaged China, South Korea, and Kazakhstan in particular, financing capacity-building projects to help these countries to develop their emissions trading system. Moreover, since the early 2010s, the EU has used regional diplomatic forums, such as the EU–Africa Partnership (launched in 2007 to increase cooperation with the African Union and its members) and the EU–Community of Latin

Table 4.4 Main EU bilateral initiatives and dialogues in the area of climate change, energy, and renewable energy (2005–2020)

Country	Year: initiative/dialogue
Brazil	2007: EU–Brazil Strategic Partnerships (includes dialogue on energy issues)
Canada	2007: High Level Dialogue on Energy (upgraded in 2014)
China (*)	2005: EU–China Energy Dialogue 2005: EU and China Partnership on Climate Change 2010: Joint Statement on Dialogue and Cooperation on Climate Change 2015: EU–China Joint Statement on Climate Change 2016: EU–China Roadmap on Energy Cooperation 2018: EU–China Leaders' Statement on Climate Change and Clean Energy 2018: Memorandum of Understanding on ETS 2019: Upgrading of the EU–China Energy Dialogue 2019: EU–China Energy Cooperation Platform
India	2005: EU–India Energy Panel (revised in 2008) 2012: Enhanced Cooperation on Energy between the EU and India 2016: Clean Energy and Climate Partnership 2020: EU–India Joint Declaration on Resource Efficiency and Circular Economy
Indonesia	2016: Working Group on Environment and Climate Change
Japan	2011: EU–Japan Energy Dialogue
Mexico	2008: EU Mexico Strategic Partnerships (includes dialogue on energy issues)
South Africa	2007: EU–South Africa Forum on Environment, Climate Change, and Sustainable Development (upgraded in 2016)
South Korea	2016: EU–Republic of Korea ETS cooperation 2018: EU–Republic of Korea Working Group on Energy, Environment and Climate
US	2009: EU–US Energy Council

Notes: The EU has also signed Memoranda of Understanding for energy cooperation with Kazakhstan (2006), Turkmenistan (2008), and Uzbekistan (2011); (*) = this table only mentions the main EU–China agreements
Sources: Author's elaboration from EU website

American and Caribbean States (EU–CELAC, established in 2011), and its enhanced partnership with the ASEAN (since 2013), to promote its views on the energy transition in the Global South. These diplomatic engagements, in turn, have helped the EU to reinforce its role in global multilateral climate negotiations.

At the multilateral level, the EU has also increased its involvement in international (energy) organizations (i.e., the IEA and IRENA) and forums (the G7 and G20) supporting the energy transition and ambitious climate actions. The Commission is also part of high-level global forums for green technology cooperation and development, such as the Clean Energy Ministerial, and Mission Innovation.

Finally, another component of the 'external face' of the EU green regulatory state manifested in the trade and investment issue area. In 2011, the Commission proposed extending the mandate of the Energy Charter Treaty to renewables, but without success. The inability to incorporate a green dimension within this treaty resulted (in 2023) in the Commission's proposal of a coordinated EU withdrawal from it. However, since the early 2010s, the EU has begun to include sustainable development chapters in its bilateral trade deals. This move, implemented by DG TRADE, was supported by several European non-governmental organizations involved in environmental protection, climate change, and development cooperation. However, it also received a positive response from EU policymakers, who were willing to enlarge the armamentarium of the bloc's green external action. The first of this new generation of trade agreements was signed with South Korea, and further such trade agreements were subsequently signed with several other countries in Asia, Latin and Central America, and North America (Table 4.5). These agreements incorporate chapters that refer to promoting the implementation of multilateral agreements on climate change (i.e., the UNFCCC, the Kyoto Protocol, and, later, the Paris Agreement) and on the development of renewable energy (Velut et al. 2022). For renewable energy (covered directly under sustainable development chapters or placed under energy and raw materials chapters), these deals include provisions to facilitate the removal of barriers to trade and investments in related goods and services and for regulatory cooperation. It is worth noting that provisions on the Paris Agreement and on renewable energy are also included in the EU–UK Trade and Cooperation Agreement, which entered into force in 2021 to govern the relationship between the bloc and the UK after Brexit. Due to Brexit, the UK is no longer part of the EU internal energy market and EU ETS, which was replaced by the UK Emissions Trading Scheme in 2021.

The EU also proposed a similar sustainable development chapter in its negotiations with the US for the Transatlantic Trade and Investment Partnership, a project that was abandoned during the Trump administration, however (see Chapter 5). In 2014, the EU—together with other WTO members, including the US—launched an initiative for a new multilateral agreement under the WTO to liberalize trade in environmental goods (i.e., the Environmental Goods Agreement, EGA). The EU regarded this agreement as an important step to increase the coherence between the global trade and climate change regimes. However, the EGA talks, which were already very difficult owing to the different positions among major powers, have been stalled since 2016. This outcome has created a further incentive for the EU to pursue its goals following the 'regulation through trade' logic. In practice, however, it is not easy for the EU to sign such deals, many of which (as at 2023) were still awaiting ratification or were in the negotiation stage (Table 4.5). This is also the case with the EU–China Comprehensive Agreement on Investment, which has been gradually discarded since 2022, as was anticipated. Moreover, these EU bilateral deals envisage only loose enforcement mechanisms (Velut et al. 2022); they do not include provisions on trade remedies and/or sanctions, such as compensation, for non-compliance with or failure to implement sustainable development chapters. In other words, these trade deals represent a soft version of green regulatory power. They are primarily instrumental in creating additional venues for dialogue and policy exchanges with third states; this includes the involvement of civil society organizations, which take part in the discussion and monitoring through dedicated forums or advisory groups. It was only after the EGD that the EU began sharpening this component of its green regulatory power strategy (see below).

Table 4.5 EU trade deals with sustainable development chapters and references to climate change and renewable energy provisions (2011–2020)

In force	Awaiting ratification (*)	Ongoing negotiations (*)
2011: EU–Republic of Korea Free Trade Agreement 2012: EU–Central American (**) Association Agreement 2013: EU–Colombia Comprehensive Trade Agreement 2013: EU–Peru Comprehensive Trade Agreement 2017: EU–Ecuador Comprehensive Trade Agreement 2017: EU–Canada Comprehensive Economic and Trade Agreement (***) 2019: EU–Japan Economic Partnership Agreement 2019: The EU–Singapore Trade and Investment Protection Agreements 2020: EU–Vietnam Trade an Investment Protection Agreement	2018: EU–Mexico Association Agreement 2019: EU–Mercosur Comprehensive Trade Agreement 2020: EU–China Comprehensive Agreement on Investment	2016: EU–Indonesia Trade Agreement 2018: EU–Australia Trade Agreement

Notes: (*) as of 2023; (**) Central American = Costa Rica, El Salvador, Guatemala, Honduras, Nicaragua, Panama; (***) = the agreement with Canada does not include explicit references to the UNFCCC and the Kyoto Protocol
Sources: Author's elaboration from EU websites and Velut et al. (2022)

Going beyond the EU regulatory state

After the 2008 economic crisis, and in parallel with the greening of the EU regulatory state, the Commission increased its efforts to boost investment in energy networks and low-carbon technology on the continent. This more interventionist turn developed in connection with the transformation and greening of the EU innovation and industrial policy, which gradually departed from the previous horizontal approach to embrace a more sector-specific, mission-oriented, and strategic approach supported by new means and a stronger focus on renewables and key low-carbon technology (Figure 4.3). This trend was initiated under the second Barroso Commission (2009–2014) and was subsequently strengthened under the Juncker Commission (2014–2019), which expanded the set of instruments deployed by the EU, including industry alliances and financial tools in closer cooperation with the EIB, and linked EU energy, climate, and innovation policy more effectively. Finally, this trend was further improved under the von der Leyen Commission (2019–2024), in a context of growing international competition and risks of excessive dependencies in critical sectors for the green (and digital) transition that required additional tools for fostering the EU's open strategic autonomy and supporting European business (Figure 4.3). These included screening measures for foreign direct investments and takeovers—an EU framework for investment screening was established with Regulation (EU) 2019/452 (effective since 2020)—instruments for analysing and addressing risks in strategic value chains, selected upgrading of industrial capacity to reduce dependencies, such as those in the area of renewables and low-carbon technology (from PV and wind equipment to batteries and electrolysers), and a 'more strategic approach to international standardization activities' (European Commission 2022d, 6). This latter point was underlined by the 2022 Communication, An EU Strategy on Standardisation, which linked Europe's technological sovereignty to the bloc's ability to be a global standard-setter in the green and digital sectors. It is worth recalling that after the Covid-19 crisis, the RRF also provided member states with access to loans and grants to support green investments, including for the decarbonization of industry (also through the establishment of 'hydrogen valleys') and the strengthening of low-carbon supply chains (Tagliapietra, Trasi and Veugelers 2023). Finally, in the wake of the Russian invasion of Ukraine and the launch of the REPowerEU Plan, the Commission additionally fostered the integration of the EU industrial, energy and climate policy, as exemplified by the 2023 Communication A Green Deal Industrial Plan for the Net-Zero Age and related initiatives such as the NZIA and the CRMA regulations, operative from 2024 (see below).

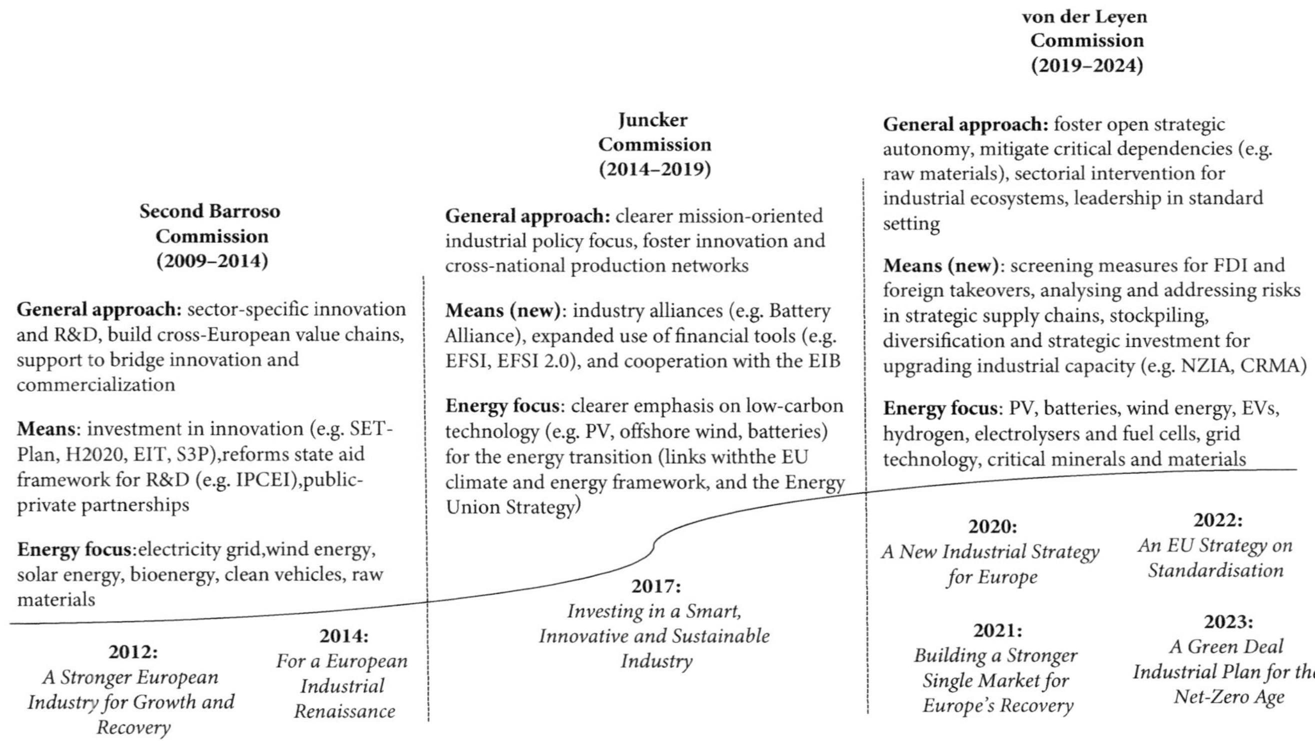

Figure 4.3 Upgrading and greening the EU's innovation and industrial policy

Notes: Italics indicate the main Communications on industrial policy issued by the European Commission in each period
Source: Author's elaboration based on Di Carlo and Schmitz (2023), and EU policy documents (see also Figure 4.2, Section B, and the main text for additional details)

As anticipated, however, this more interventionist turn that expanded the EU's green actions well beyond the narrow market-liberal focus of the regulatory state took the form suggested by the catalytic state perspective (Prontera and Quitzow 2022). This perspective resembles the neo-developmental (network) state approach as the catalytic state mimics its core functions (e.g., Di Carlo and Schmitz 2023). However, this perspective recognizes that a similar model adopts specific features in the context of the EU due to its (institutionally) constrained nature as a quasi-state—with a very limited budget and independent revenue-raising capability—and its partial competences in the areas of energy, industrial, and innovation (R&D) policy. Indeed, in the case of R&D, the competences are shared between the EU and its member states; whereas in the industry domain the EU has limited influence, as it is only competent to support and coordinate national efforts. Moreover, unlike its nation state counterparts, the Commission does not have obvious EU champions to nurture and support for global competition. Rather, the European industrial landscape is occupied by several national companies, which have more connections with the national governments (often of the larger EU member states) than the EU institutions. Therefore, in 2017, EU member states rejected French President Emmanuel Macron's proposal for creating an EU-level innovation agency modelled after the famous US Defence Advanced Research Projects Agency (DARPA; see Chapter 5), as this approach was considered too top-down and centralized (Mocanu and Thiemann 2023). Consequently, as a catalytic state, the EU has sought to steer and upgrade European green innovation, industrial, and technological capacity and to promote energy networks in a more indirect way (and with fewer resources and legal competences) than a traditional state actor. The Commission has acted as a catalyst and facilitator of investment projects, in particular. It has achieved this by providing supplemental resources, expertise, and incentives to national actors, promoting the creation of multi-stakeholder platforms, public–private partnerships, and pan-European industry alliances (e.g., the European Battery Alliance, the European Raw Materials Alliance, the Clean Hydrogen Alliance, the Solar PV Alliance, which are managed by the Directorate-General for Internal Market, Industry, Entrepreneurship and SMEs, DG GROW) to increase coordination between industrial actors and EU- and national-level agencies involved in R&D, and by deploying financial instruments, such as blending and guarantees, to leverage private sector funding. With regard to such leverage of private sector funding, in particular, the EU has increasingly mobilized the European development banks—the EIB as well as the EBRD—to catalyse investment in strategic projects, often in cooperation with the national development banks of the EU member states (e.g., Mertens and Thiemann 2018, 2019; see also Lepont and Thiemann 2024).

The first steps towards the catalytic state were taken between 2008 and 2013 with the European Economic Recovery Plan (EERP) and other measures targeting energy infrastructure (e.g., the Trans-European Networks for Energy, TEN-E, and the Connecting Europe Facility for Energy, CEF-E) and low-carbon innovations (e.g., SET Plan, NER 300, FP7-JUs, EIT-InnoEnergy; Figure 4.2, Section B). While the EU had indeed already implemented programmes for energy R&D and infrastructure in previous years (e.g., Eikeland and Skjærseth 2023), the scale, scope, and green focus of these actions had been widely augmented since then. This shift was institutionalized under the Juncker Commission, which in response to the slow recovery after the crisis and increasing geopolitical (Russia) and geoeconomic (China) challenges, launched the Investment Plan for Europe and the European Fund for Strategic Investments (EFSI) in 2014–2015; the so-called Juncker Plan. This plan aimed at mobilizing €500 billion in investment over five years, primarily through the use of guarantees and blending (Mertens and Thiemann 2019). These new vehicles significantly increased the ambition of the EU's green investment agenda and expanded the use of new forms of financial support involving EU development banks and private–public partnerships for promoting low-carbon technology and energy networks. This was paralleled with several other initiatives for innovation and deployment that increasingly targeted renewables and low-carbon technology, such as PV, offshore wind, and batteries (Figure 4.2, Section B). Moreover, in order to foster investment in strategic industrial projects, with the Important Projects of Common European Interest (IPCEI), the Commission has since 2014 allowed the targeted used of state aid by member states, thus overcoming one important limit that had traditionally constrained more active industrial policy across Europe. Finally, the EU catalytic state was further expanded and institutionalized under the von der Leyen Commission when, as illustrated, the EU industrial turn was more clearly linked to the new goals of open strategic autonomy and the EU's energy transition and decarbonization objectives (Figure 4.2, section B; see also Figure 4.3).

More importantly, following the same internal–external path dependencies which drove the externalization of the EU green regulatory state, these changes in state–market relations affected the external action of the EU during the 2010s, which shifted from a prevalent focus on regulatory power towards catalytic power strategies (Prontera 2019; Prontera and Quitzow 2023). This change also reflected an awareness of the limited effectiveness of the export of the EU's rules and preferences—apart from the very close neighbourhood—and concerns about losing influence vis-à-vis other major powers in important regions of the world. The 2013 TEN-E scheme already provided for supporting investments in energy infrastructure projects involving third states, in particular neighbouring countries. At that time, its main focus was on diversification of gas supply, notably in the context of the so-called Southern Gas Corridor. Since 2015, however, the EU has exported this system of planning and incentives to the EnC countries. They adopted the 2013 TEN-E scheme (the same scheme applies to the EEA countries), which was instrumental in promoting investment in new cross-border electricity projects in the Western Balkans and Eastern Europe. Subsequently, the 2022 TEN-E and CEF-E review further expanded the possibility of granting political and financial support for electricity, renewable energy, and hydrogen cross-border projects involving third countries. In 2022, under the CEF-E, the EU approved the first grant (€308 million) for a cross-border electricity infrastructure developed by an EU member state and a country outside the EU: the ELMED interconnector project, a subsea high-voltage power cable to connect Italy with Tunisia, promote the integration of a Euro-Mediterranean electric grid, and foster the deployment of renewable energy in North Africa. Similarly, under the CEF-E, the EU has provided support for the EuroAsia Interconnector, another subsea high-voltage power cable intended to link Greece, Cyprus, and Israel. In line with the catalytic power perspective, in both cases, the EU funding and the diplomatic backing by the Commission assisted in creating multi-actors coalitions—composed of governments and companies—and mobilizing the resources needed for the implementation of the projects.

In addition, the EU has over time established a system of so-called blending facilities for promoting investments in the area of renewable energy and electricity networks in the EU's neighbourhood and in the Global South. Blending, in this context, refers to the combination of EU grants with loans or equity from public and/or private financiers. It helps the EU to mobilize additional resources, primarily from European development financial institutions (DFIs)—the EIB, the EBRD, and the development agencies from the member states—as well as from regional development banks and public and private partners in third states. The increase in blending is indeed part of a more general increase of EU development financing, aimed at enhancing the influence and impact of EU external action in the wake of a changing international environment (Prontera and Quitzow 2023). The EU's official development assistance more than doubled between the mid-2000s and the mid-2010s, increasing from approximately €8 billion to more than €16 billion, and has remained relatively stable over the decade (ibid.). This has gone in tandem with experimentation with new forms of aid delivery. Initially, the EU significantly increased the use of so-called budget support, which represents a direct transfer of funds to a partner country's treasury. Between 2014 and 2021, the EU's annual disbursements for budget support averaged approximately €1.7 billion, geographically focused primarily on sub-Saharan Africa (35%) and the EU's neighbourhood (31%), followed by Asia (16%) and Latin America (6%; European Commission 2022e). This type of aid is not linked to any specific investment programme, but allows the EU to engage with developing countries in a dialogue at the policy level. Since the approval of the 2015 UN Sustainable Development Goals and the Paris Agreement, the EU has used this channel to promote climate and sustainable development objectives abroad, including measures and reforms for renewable energy deployment. Although this strategy relies on the use of financial incentives, primarily in the form of technical assistance, it is closer to a manifestation of green soft power. By engaging in policy dialogue and supporting certain reforms and objectives the EU aims at shaping the milieu in which the global energy transition takes places. These bilateral efforts run in parallel with the EU backing for global multilateral initiatives for dialogue and cooperation on climate change, such as the Global Climate Change Alliance launched by the European Commission in 2007 and upgraded in 2015. The EU provided approximately €730 million for this initiative in the 2007–2020 period (whereas contributions for multilateral initiatives such as the Green Climate Fund are mainly provided by EU member states).

However, only a small proportion of the EU budget support (nearly 3%) has been specifically allocated to the promotion of renewable energy projects abroad (European Commission 2022e). The EU has generally deployed other financial tools with a view to achieving this objective. Indeed, at the same time that the EU started scaling up budget support, it also began experimenting with blended finance. Unlike budget support, this approach resembles the catalytic green power strategy, in that it aims at creating multi-actor coalitions to achieve the EU goals by combining economic incentives to mobilize the resources of others and modes of network diplomacy to support investments and projects in third countries. Originally, the EU's approach to blended finance emerged as a niche within the EU development policy. Over time, however, it evolved to become an important instrument for the wider EU green foreign energy policy. This change began to take shape during the EU's 2007–2014 multi-annual budget (Mah 2018). In 2007, building on the Joint Africa–EU Strategy, the European Commission—in cooperation with a number of EU member states—launched a regional blending facility for sub-Saharan Africa, the Africa–EU Infrastructure Trust Fund (EU–AITF). This facility, managed by the EIB, provided financial resources (primarily in the form of grants for investments) specifically for electricity networks and renewable energy projects. From 2007 to 2019, thanks to the EU–AITF, the cooperation between the EU, member states, and regional DFIs, and the involvement of (primarily) public and/or private partners in third countries, the EU has supported projects in the wind and solar sector for the provision of approximately 1,100 MW of renewable installed capacity (EU-AITF 2020).

This new approach was expanded in the wake of the global economic crisis. In 2008–2009, two regional blending facilities were established: one for the Western Balkans (the Western Balkans Investment Framework, WBIF) and one for the ENP countries, namely the Neighbourhood Investment Facility (NIF; Table 4.6). Then, from 2010 to 2012, other (smaller) blending facilities were set up for all the geographical areas covered by the EU development policy, from Latin and Central America to the Asia Pacific (Table 4.6). Unlike the EU–AITF, these new blending facilities—funded by the existing budget instruments for EU external action—were managed by the Commission, in particular DG NEAR and the Directorate-General for International Cooperation and Development (DG DEVCO; renamed the Directorate-General for International Partnerships, DG INTPA, in 2021; Table 4.6). This innovation increased the Commission's capacity to politically steer these economic instruments. However, these blending facilities responded to the same logic: supporting investment projects abroad by mobilizing resources from EU and non-EU DFIs, as well as public and private partners, in coordination with recipient countries. Through the governance system of the blending facilities, in particular—which also involves the EU member states, the EIB and other DFIs—the European Commission can structure cooperation and create diplomatic networks to achieve its objectives and increase its influence (Prontera and Quitzow 2023). In the 2014–2020 EU multi-annual budget, the funding allocated for blending was drastically increased (from approximately €1.6 billion to more than €7 billion). Still, the EU geographical focus remained on the EU's neighbourhood and Africa, while the EIB, the EBRD, the German KfW, and the French AFD continued to be the main DFIs involved in blending operations (Table 4.6). Renewable energy and climate action occupied a stronger position in these operations, reflecting the evolving global development and climate agenda (e.g., the 2030 UN Agenda for Sustainable Development and Paris Agreement), to which the EU itself had contributed, as well the growing emphasis of the EU on the energy transition in the wake of the Energy Union Strategy and the 2030 Climate and Energy

Framework. In 2015, in the Western Balkans, the Commission also pledged an additional €1 billion in EU grants for a Connectivity Agenda that aimed at improving transport and energy infrastructures, as well as renewable energy projects. Indeed, the countries in this region involved in the EU's enlargement policy and the EnC were facing the challenging task of meeting the obligations arising from the transposition of EU energy and climate legislation and those assumed under the Paris Agreement. Nevertheless, the Connectivity Agenda served to better integrate these countries with the EU energy market and to counter China's growing influence in the Western Balkans. Overall, approximately 500 MW of wind and solar capacity were installed between 2007 and 2020 under the WBIF (WBIF 2021). Moreover, important cross-border electricity projects—such as the Trans-Balkan Electricity Corridor that links Serbia, Montenegro, and Bosnia and Herzegovina with a high-voltage electricity line—were launched.

In 2016–2017, the EU issued the External Investment Plan and its financial arm, the European Fund for Sustainable Development (EFSD). This initiative was the 'external face' of the Juncker Plan, on which it built. Regulation 2017/1601 establishing the EFSD provided for additional resources for blending (up to €2.6 billion) and a new EFSD Guarantee of €1.5 billion to better mobilize private sector funding. It also provided that approximately 30% of the EFSD funding should contribute to climate action, renewable energy deployment, and resource efficiency (the EFSD Guarantee, in particular, had a strong focus on renewable energy projects). Once again, the EIB played an important role in implementing these goals: from 2018 to 2021 it committed approximately €3 billion for renewable energy projects abroad, especially in Africa and Latin America (EIB 2023). Moreover, the EFSD provided for the adoption of a more holistic approach to the use of financial support, which was also intended as an entry point for policy dialogue and reforms. Finally, the EFSD improved and centralized the governance framework for the EU blending facilities. The facilities for the EU's neighbourhood and Africa—renamed the Neighbourhood Investment Platform (NIP) and Africa Investment Platform (AIP)—were placed under the EFSD. Moreover, the EEAS was involved, together with the Commission, in the management of the EFSD with a view to pursuing EU foreign policy objectives more effectively by means of these financial tools. Indeed, the External Investment Plan was also guided by geopolitical goals; specifically countering the rising influence of China in Africa and the EU's neighbourhood.

The shift of blending (and guarantees) from the niche of development cooperation towards the wider machinery of EU foreign policy continued under the von der Leyen Commission. In 2021, the EFSD was upgraded

Table 4.6 The EU regional blending facilities: funding instruments, resources, actors, and renewable energy and electricity projects (2007–2020)

Region	Blending facility	Years	Managed by/ EU funding instrument (ç)	Resources allocated (million €)		Main DFIs involved (^)	% allocated to energy projects	Energy projects (main sectors)
				2007–2014	2014–2020			
Western Balkans	Western Balkans Investment Framework (WBIF)	2009–2020	DG NEAR/IPA	301	1170 1000 (*)	EIB, EBRD, KfW, CEB, AFD	30%	Wind, solar, electricity network, hydro (minor), natural gas (minor)
EU Neighbourhood (ENP countries)	Neighbourhood Investment Facility (NIF)	2008–2015	DG NEAR/ENPI–ENI	648	2273	EIB, EBRD, AFD, KfW	33%	Renewable energy, including wind, solar and hydro (minor), electricity networks, natural gas (minor)
	Neighbourhood Investment Platform (NIP)	2017–2020	DG NEAR–EEAS/EFSD			EIB, EBRD, AFD, KfW	18%	Renewable energy, including wind, solar and hydro (minor), electricity networks
Sub-Sahara Africa	Africa-EU Infrastructure Fund (AITF)	2007–2019	EIB/EDF	366	2487	EIB, EBRD, AFD, KfW, AfDB	67%	Renewable energy, including wind, solar and hydro (minor), electricity networks
	Africa Investment Facility (AIF)	2015–2016	DG DEVCO/DCI–EDF				30%	
	Africa Investment Platform (AIP)	2017–2020	DG DEVCO –EEAS/EFSD				27%	

Continued

Table 4.6 *Continued*

Region	Blending facility	Years	Managed by/ EU funding instrument (ç)	Resources allocated (million €)		Main DFIs involved (^)	% allo-cated to energy projects	Energy projects (main sectors)
				2007–2014	2014–2020			
Latin America and Caribbean	Latin America Investment Facility (LAIF)	2010–2020	DG DEVCO /DCI	230	475	EIB, AFD, KfW, AECID, CAF, IDB	27%	Renewable energy, including wind, solar and hydro (minor), electricity networks
	Caribbean Investment Facility (CIF)	2012–2020	DG DEVCO /EDF			EIB, AFD, CDB, IDB	50%	
Asia-Pacific	Asia Investment Facility (AIF)	2012–2020	DG DEVCO /DCI	104	733	EIB, EBRD, AFD, KfW, ADB	13%	Renewable energy, including wind, solar and hydro (minor), electricity networks
	Investment Facility for Central Asia (IFCA)	2010–2020	DG DEVCO /DCI			EIB, EBRD, AFD, KfW, ADB	20%	
	Investment Facility for the Pacific (IFP)	2012–2020	DG DEVCO /EDF			EIB, EBRD, AFD, KfW, ADB	30%	

Notes: (ç) = Instrument for Pre-Accession (IPA); European Neighbourhood and Partnership Instrument (ENPI); European Neighbourhood Instrument (ENI); Development Cooperation Instrument (DCI); European Development Fund (EDF); European Fund for Sustainable Development (EFSD); (*) = resources committed under the 'Connectivity Agenda'; (^) = for each blending facility the EU issues a list of eligible development financial institutions (DFIs) that usually includes European and regional development banks. DFIs: European Investment Bank (EIB); European Bank for Reconstruction and Development (EBRD); Kreditanstalt für Wiederaufbau (KfW, Germany); Agence Française de Développement (AFD, France); Council of Europe Development Bank (CEB); African Development Bank (AfDB); Agencia Española de Cooperación Internacional Para el Desarrollo (AECID, Spain); Inter-America Development Bank (IDB); Development Bank of Latin America and the Caribbean (CAF); Asian Development Bank (ADB)

Sources: Author's elaboration from the blending facilities Operational Reports accessed through the EU websites

to the EFSD+, which aggregated under a common budget instrument and governance framework all the existing regional blending facilities, with the exception of the WBIF funded by the Instrument for Pre-accession Assistance (IPA). The EFSD+—backed by a new External Action Guarantee of €10 billion from the EU budget (to cover operations up to approximately €50 billion for the period 2021–2027)—became the financial arm of the NDICI-Global Europe. This economic tool is aimed at contributing to achieving the international commitments that the EU has agreed to, in particular the 2030 Agenda and its Sustainable Development Goals, as well as the Paris Agreement. However, in line with the so-called policy first approach, it has placed a stronger emphasis on the EU's strategic objectives and pursues foreign policy priorities, such as those of the Global Gateway Strategy.

Greening the external action of the EU

As mentioned, with the EGD, the EU increased its efforts to further upgrade its green posture and improve the coherence of its different elements. Since 2020–2021, the implementation of the (evolving) EGD agenda has manifested in several interconnected sectors of the EU's external action (Figure 4.4). First, the EGD was embedded in the EU's energy and climate diplomacy. With the 2020 Hydrogen Strategy, green hydrogen was then mainstreamed in the EU's international, regional, and bilateral energy engagements. Later, the EGD permeated the overall EU energy policy, which was transformed in the wake of the Russian invasion of Ukraine and the launch of the REPowerEU Plan, as well as the EU trade and industrial policy (including in the area of standardization), and the initiatives for critical raw materials (see also Table 4.8). The EGD was also central in the formulation of the Global Gateway Strategy, backed by the new financial tools for the EU's external action (i.e., the NDICI-Global Europe and the EFSD+) and sustained by the so-called Team Europe approach. This latter approach emerged as a response to the Covid-19 crisis to bring together the EU and the member states with their financial and development institutions. It was then translated into the bloc's foreign green engagements. A similar cooperation, as explained, had already occurred for the implementation of the EU blending facilities. However, under the Team Europe initiatives it was further institutionalized and upgraded. Moreover, the EGD evolving agenda was reflected in the new regional strategies and (bilateral) strategic alliances and partnerships launched since 2020 (Tables 4.7, 4.9, and 4.10). Finally, as part of the EGD, the EU issued its CBAM, the world's first carbon border tariff, with

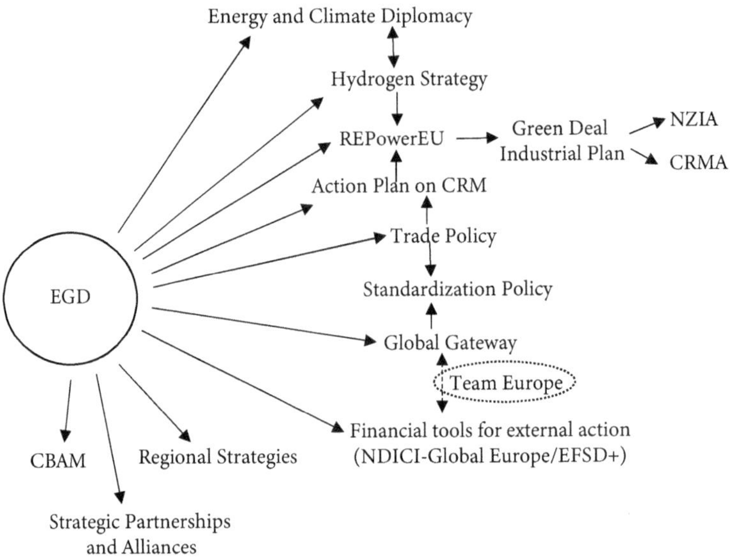

Figure 4.4 The ramification of the European Green Deal (EGD) agenda in the EU external action
Source: Author's elaboration

potentially an important (global) external reach. All these initiatives, which were closely linked, have been affected not only by the EGD but also by the new 'geopolitical' turn taken by the EU: these two developments combined to transform the EU green foreign energy policy and green power strategy.

Regarding energy and climate diplomacy, the EGD agenda influenced both the bilateral and multilateral actions of the EU. Bilaterally, the EU sharpened its focus on promoting the energy transition in third countries, especially among its main suppliers of fossil fuels. Recognizing that these countries would be adversely affected by the reduction in exports to Europe, the EU aimed to support their economic and political stability, which is highly dependent on oil and gas revenues. This goal is particularly important for producer states such as Algeria, with whom the EU has a strategic energy partnership since 2013–2015, or Azerbaijan (a new strategic partnership covering renewables with this country was signed in 2022; see Table 4.7). On the other hand, as illustrated, the adoption of the EU Climate Law and the Fit for 55 Package were important for supporting EU multilateral climate diplomacy in the run up to the 2021 COP 26 in Glasgow. At 2022 COP 27, in Egypt, the EU confirmed its role as 'lediator' in multilateral negotiations, facilitating the agreement that provided for the new 'loss and damage'

Table 4.7 New EU bilateral initiatives and dialogues in the area of renewable energy, low-carbon technology, and raw materials (2021–2023)

Year	Initiative/dialogue
2021	EU–Canada Strategic Partnership on Raw Materials
2021	EU–Ukraine Strategic Partnership on Raw Materials and Batteries
2021	EU–Japan Green Alliance
2021	EU–US Trade and Technology Council
2022	EU–Kazakhstan Strategic Partnership on Raw Materials, Batteries and Renewable Hydrogen
2022	EU–US Task Force on Energy Security (LNG, climate change, renewable energy)
2022	EU–Azerbaijan Strategic Partnership in the Field of Energy (gas, renewable energy, climate change)
2022	EU–Egypt–Israel Memorandum of Understanding (LNG, climate change, renewable energy, hydrogen)
2022	EU–Egypt Strategic Partnership on Renewable Hydrogen
2022	EU–Morocco Green Partnership (renewable energy, climate, green hydrogen)
2022	EU–Namibia Partnership on Raw Materials and Renewable Hydrogen
2022	EU–India Trade and Technology Council
2023	EU–Norway Green Alliance

Sources: Author's elaboration from EU websites

fund. In addition, the EU, in cooperation with the US and other industrialized countries, including the UK, Canada, and Japan, helped reinforce commitments to reduce methane emissions and expand the membership of the Global Methane Pledge initiative, already established at the COP 26. Sim ilarly, the EU climate diplomacy contributed to the agreement reached at the 2023 COP 28 in Dubai concerning the 'transition away from fossil fuels', as well as to the Global Renewables and Energy Efficiency Pledge, which aims to triple the world's installed renewable energy generation capacity by 2030. At Dubai, the EU, along with Germany, also worked to expand the membership of the Climate Club, a (still embryonic) concept endorsed by the US and the G7, to increase cooperation among countries more ambitious in the area of industry's decarbonization and emissions reduction (other countries involved in this initiative are Argentina, Australia, Canada, Colombia, Egypt, Indonesia, Japan, Kenya, Mozambique, Morocco, Ukraine, and the UK).

Along with its member states, and working with the G7 partners, the EU was also active in promoting abroad the 'just transition' agenda, which was strengthened domestically with the EGD. At COP 26, the EU, along with

France, Germany, the US, and the UK, launched the first Just Energy Transition Partnership with South Africa, backed by an initial commitment of $8.5 billion. This partnership aimed at helping the country accelerate the phasing out of coal and the decarbonization of the power sector through the promotion of renewables while ensuring social measures for affected workers and communities. In 2022–2023, the EU participated in the launch of three other Just Energy Transition Partnerships (JETPs), backed by financial support and capacity building programmes, with Vietnam, Indonesia, and Senegal. These were developed in cooperation with the US and other industrialized countries (i.e., the UK, France, Germany, Italy, Canada, Japan, Norway, and Denmark) gathered under the International Partners Group, an initiative originated within the G7 and linked to the 2022 Partnership for Global Infrastructure and Investment (the evolution of the 2021 US-led Build Back Better World; see Chapter 5). The Just Energy Transition Partnership with Indonesia aimed at mobilizing $20 billion. Via the EIB, the EU committed €1 billion for the decarbonization of the Indonesian power sector through renewable energy. This was the biggest single climate initiative ever funded by the EU (Goldthau and Youngs 2023).

In the field of green hydrogen, the 2020 strategy outlined specific targets by 2030 for electrolyser capacity and production (Table 4.8). It also emphasized the integration of hydrogen into the EU's external action. Through the European Commission, the EU already participated in the major multilateral bodies promoting R&D and cooperation on hydrogen—such as Mission Innovation and the International Partnership for Hydrogen and Fuel Cells in the Economy—as well as in the works of the G7 and the G20 on the matter. However, the new strategy adopted a sharper approach to foster cooperation, especially with North African countries seen as potential suppliers of green hydrogen to Europe. Indeed, unlike in the case of China and the US, the EU's strategy for hydrogen has a stronger outward-looking dimension, aiming to leverage the renewable energy potential of neighbouring regions and address the EU's limitations in this regard. The EU's approach also focused on the promotion of international standards for the nascent hydrogen industry and trade, including the elaboration of sustainability criteria for its supply chain. For this, the European Commission utilized the CertifHy initiative, a multi-stakeholder platform financed by the Clean Hydrogen Partnership, to promote certification schemes for green hydrogen aligning with the EU regulatory framework (particularly the EU directives on renewables). The 2022 REPowerEU Plan then set a target of 10 million tonnes of green hydrogen from imports by 2030 (Table 4.8). This required improving the foreign policy dimension of the EU hydrogen strategy by supporting the development of three major hydrogen import corridors—via the Mediterranean Sea, the North Sea region (i.e., Norway and the UK) and (potentially) Ukraine—and through bilateral partnerships to foster green hydrogen production in third states, such as Morocco, Egypt, Kazakhstan, and Namibia (Table 4.7). Importantly, with the EU's increased activism to diversify gas supplies post the Russian invasion, the bloc linked gas cooperation with long-term cooperation on renewable energy and green hydrogen. This linkage was evident in partnerships with the US, Azerbaijan, Egypt, and Israel (Table 4.7). Promoting hydrogen abroad is also considered by the EU to be instrumental in avoiding stranded assets and helping decarbonization in third states (green hydrogen could possibly be imported through the existing gas pipelines). This connection between gas and hydrogen was confirmed with the establishment of the EU Energy Platform (operative from 2023). This tool created an EU-level (voluntary) system for demand aggregation and joint purchasing of natural gas and liquefied natural gas (LNG) in order to leverage the bloc's market power vis-à-vis potential suppliers and could be extended to hydrogen in the future.

To reduce dependency on Russian fossil fuels, the REPowerEU Plan also increased the targets for renewable energy installed capacity by 2030, especially in the solar sector, where it provided to more than double the EU capacity. Moreover, this plan, and the related External Energy Strategy, further placed the energy transition at the centre of the EU international energy policy in connection with other EGD initiatives and foreign policy tools (Table 4.8, Figure 4.4; see also below). Some of these initiatives, in the area of industrial policy and critical minerals, were additionally developed with the NZIA and CRMA that operationalized the Green Deal Industrial Plan issued on February 2023 (Figure 4.4; Table 4.8). The EU, in particular, included a focus on improving the resilience and sustainability of critical supply chains through multilateral, bilateral, and unilateral trade measures. Through the 2021 Trade Policy Review, the EU reasserted its commitment to the CBAM. It also proposed the establishment of an Anti-Coercion Instrument (ACI) to enable the EU to respond to economic coercion exercised by third states and better defend its interests on the global stage. Both of these new (unilateral) tools became operative at the end of 2023. Moreover, the 2021 Trade Policy Review called for strengthening the enforcement of the sustainability dimension of the bloc's trade agreements, i.e., the 'regulation through trade' approach. In 2022, the EU reached a new deal with New Zealand that, unlike the previous ones, allowed for the possibility of sanctions as a last resort in the case of serious violations of the Paris Agreement. In line with the 2020 Action Plan on Critical Raw Materials, the EU began using its trade deals

Table 4.8 EU initiatives in the area of energy policy, trade and industrial policy, and critical raw materials and their green foreign energy policy dimension (2020–2023)

Initiative (Year)	Key objectives for selected renewable technology and low-carbon supply chains	Green foreign energy policy dimension
Energy policy		
Hydrogen Strategy (2020)	Target by 2030: • 40 GW of green hydrogen electrolysers installed • Production of up to 10 million tonnes of green hydrogen	• Mainstreaming hydrogen in the EU's international, regional, and bilateral energy and diplomatic efforts • Promote cooperation with the European Neighbourhood countries from research and innovation to regulatory framework, increase investments and hydrogen trade (focus on North Africa as a potential supplier of renewable hydrogen to the EU) • Promote standards and sustainability criteria for hydrogen in multilateral fora
REPowerEU Plan (2022)—External Energy Strategy (2022)	Target by 2030: • 1236 GW of renewable energy generation capacity • 600 GW of solar PV • 20 million tonnes of green hydrogen (of which 10 million tonnes of imports)	• Accelerate the global green and just energy transition (combing financial support, reforms, and capacity building) • Promote hydrogen import corridors via the Mediterranean Sea, the North Sea area (Norway and UK), and Ukraine • Promote green hydrogen partnerships and supporting the decarbonization in the partner countries • EU Energy Platform for the (voluntary) common purchase of gas, LNG, and hydrogen • Combine gas cooperation with long-term energy cooperation on hydrogen, and other renewable energy sources to avoid stranded assets • Support green investments, connectivity, and just transition through the Global Gateway, EU financial tools, and collaboration with multilateral organizations (e.g., IEA, IRENA) • Establishing raw materials value chains partnerships and use EU economic and trade tools to ensure undistorted access to the international markets

Trade and industrial policy		
Trade Policy Review (2021)		• Support the green transition and strengthen the resilience and sustainability of critical supply chains, multilaterally (engaging the WTO), bilaterally, and unilaterally) • Improve the effective implementation and enforcement of sustainable development chapters in trade agreements • Unilateral measures to tackle carbon leakage (i.e., CBAM) • Regulatory instrument to protect the EU from third countries' coercive actions (e.g., Anti-Coercion Instrument) • EU strategy for export credit to support EU green business and green projects abroad
EU Strategy on Standardisation (2022)	Develop standards to support: • The recycling of critical raw materials • The roll-out of the green hydrogen value chain	• Adopt a more strategic approach to international standards (focus on the green and digital transition) • Leverage trade agreements and partnerships to support shared interests in international standards setting with key partners • Create a High-Level Forum to increase coordination between the Commission and member states' standardization bodies • Leverage the Global Gateway to promote European standards
NZIA (2023)	Target by 2030: • 40% of the EU annual deployment needs by EU domestic manufacturing for solar and wind technologies, batteries, electrolysers and fuel cells, and grid technologies	• Promote industrial partnerships with like-minded countries (e.g., Net-Zero Industrial Partnerships, Trade and Technology Council) • EU export credits strategy (e.g., EU Export Credit Facility), and enhanced coordination of financial tools to support green business and enhance investments in the energy transition abroad • Work with EU–US Task Force on the IRA to maintaining and reinforcing Transatlantic low-carbon value chains • Hydrogen Bank for green hydrogen production in the EU and internationally (in coordination with the EU Energy Platform)

Continued

Table 4.8 *Continued*

Initiative (Year)	Key objectives for selected renewable technology and low-carbon supply chains	Green foreign energy policy dimension
Critical raw materials		
CRMA (2023) (*)	Target by 2030 for strategic raw materials (e.g., cobalt, copper, lithium, nickel, and REEs): • Extract 10%, process 40%, and recycle 25% of the EU annual consumption • Not rely on a single source of supply for more than 65% of the EU annual consumption	• Critical Raw Materials Club • Use free trade agreements to secure and diversify trade in critical raw materials • Promoting strategic partnerships on raw materials with a value chain approach and sustainability criteria • Use the Global Gateway to promote strategic projects along the raw materials value chain in third countries • EU Export Credit Facility for promoting investment along the raw materials value chain • Strengthening the WTO and tackling unfair trade practices related to raw materials • Promote EU standards for critical raw material-related industrial processes

Notes: (*) = the CRMA builds on the 2020 Action Plan on Raw Materials and the 2023 Commission's Communication A Secure and Sustainable Supply of Critical Raw Materials in Support of the Twin Transition
Source: Author's elaboration from EU documents and websites

to increase the resilience of its supply chains. In 2021, leveraging the EU–Canada Comprehensive Economic and Trade Agreement's mandate, notably the bilateral dialogue on raw materials, the EU launched a strategic partnership on raw materials with Canada (Table 4.7). A similar partnership was established with Ukraine in 2021, but it was halted due to the war. However, in 2022, the EU reached an agreement with Chile to modernize its 2002 Association Agreement. The new EU–Chile Advanced Framework Agreement not only envisages a sustainable development chapter but also provides for better access, investment and trade in the area of critical raw materials (such as lithium and copper) and green hydrogen. In 2023, the EU also signed a new type of agreement provided by its post-EGD green trade policy, i.e., the so-called Sustainable Investment Facilitation Agreement, with Angola, a fossil fuel exporter that holds reserves of critical minerals such as cobalt, copper, and lithium, and REEs like neodymium. This deal contains sustainable development provisions, including the Paris Agreement, as well as measures for facilitating investment and improving the country's regulatory framework.

Overall, between 2021 and 2023, the EU established several new bilateral partnerships for cooperation and dialogue with potential suppliers of critical minerals and/or green hydrogen, such as Canada, Ukraine, Azerbaijan, Egypt, Morocco, and Namibia (Table 4.7). The EU also strengthened 'green alliances' with like-minded democratic countries, such as Norway and Japan, with a strong focus on improving cooperation on regulatory matters (Norway is already an EEA country), on standards for renewables, hydrogen and low-carbon technology, as well as on sustainable finance and industry decarbonization (Table 4.7). Closer cooperation on renewable energy, low-carbon technology supply chains, green industrial policy and standards was then promoted in transatlantic relations through the 2021 EU–US Trade and Technology Council. An EU–India Trade and Technology Council was also established in 2022 to foster cooperation in the areas of green technology development, trade, investment, and standards. Renewables and climate action were also included, along with hydrogen, among the areas of cooperation of the EU–US Energy Security Task Force, which was created to help the EU reduce its gas dependency on Russia.

The 2022 EU Strategy on Standardisation also has a green foreign energy policy dimension (Table 4.8). At home, this strategy focused on the development of standards for the recycling of critical raw materials and the roll-out of green hydrogen. A Batteries Regulation was issued in 2023 (Regulation 2023/1542); it established improved rules and targets to favour the recycling of raw materials and the sustainability of batteries' supply chains. According to these new rules, which will gradually be applied from 2025, companies are called to identify, prevent and address social and environmental risks linked to the sourcing, processing, and trading of raw materials contained in their batteries, such as lithium, cobalt, and nickel. In 2023, the EU issued new detailed rules on the definition of renewable hydrogen linked to the RED III directive. In both cases, the EU aims to set global standards in these fields and support the diffusion of advanced environmental, social, and governance criteria. This goal is also reflected in the bilateral partnerships launched by the EU, such as in the case of the one established with Namibia (Table 4.7). In addition, the 2022 EU Strategy on Standardisation focused explicitly on adopting a more strategic approach, in cooperation with member states, to shape international standards by leveraging the bloc's trade agreements and partnerships and its financial engagements in the context of the Global Gateway. A High-Level Forum on European Standardisation for improving coordination between the EU and member states started operating at the beginning of 2023.

The NZIA and the CRMA are key and interconnected components of the EU rising activism in the area of industrial policy and critical minerals. The related regulations were proposed by the Commission at the beginning of 2023—with the Communication on a Green Deal Industrial Plan for the Net-Zero Age—and were agreed upon at the end of that year. Both these initiatives are based on closer links between internal and external measures and reflect the new strategic approach to renewables and low-carbon technology supply chains embraced by the von der Leyen Commission, as they aim at increasing the EU's resilience and posture vis-à-vis other major powers, particularly China (Siddi and Prandin 2023). Indeed, China is the major EU's supplier of several critical minerals and low-carbon technology (remarkably in the solar sector), and it would be impacted heavily by the targets envisaged by these initiatives (Table 4.8). The NZIA, however, was also a response to the challenges posed to the European green industry by the US IRA. A similar move was supported by the larger EU member states, especially France and Germany, which were developing analogous measures at the national level. After the Russian invasion of Ukraine and the resulting energy crisis, with the REPowerEU Plan, the Commission had already amended the RRF regulation to include an extra €72.3 billion of financing to support member states in increasing their energy security (Bocquillon, Brooks, and Maltby 2023). With these additional funds, national governments could update their National Recovery and Resilience Plans (NRRPs) by adding a chapter with reforms and investments for fossil fuel infrastructures (e.g., LNG import terminals and gas pipelines), as well as for the deployment of wind and solar energy and the expansion of manufacturing capacity for low-carbon technology and related supply chains. However, not all of the proposals of the Commission were accepted by the member states. Particularly, the Commission suggested the creation of an EU Sovereignty Fund—financed through common debt like the NextGeneration EU—to more directly support European strategic industries, expand manufacturing capacity, and address critical dependencies. Yet, this path-breaking idea was opposed by several member states, including Germany (Strupczewski 2023). In the end, it resulted only in the establishment, in 2023, of the Strategic Technologies for Europe Platform as a repackage of existing programmes and funds, in line with the catalytic state perspective, to support European low-carbon (and digital) technology.

The NZIA, however, operationalized the new strategic approach to renewables and low-carbon technology championed by the Commission and the larger member states. It set a target (by 2030) for domestic industrial capacity (40% of the EU deployment needs) for several renewable energy technologies, including solar and wind equipment (e.g., PV modules and cells, permanent magnets, blades), batteries, electrolysers and fuel cells, and grid technology (Table 4.8). Other targets were also set for carbon capture, utilisation and storage (CCUS) technology, and, on the request of the member states, nuclear energy was included among the strategic sectors supported by the NZIA. To expand manufacturing and achieve its targets, the NZIA introduced new regulatory frameworks to favour investments, mainly through regulatory sandboxes to help develop and test innovative low-carbon technologies and the so-called Net-Zero Strategic Projects. These latter projects can be implemented through accelerated and facilitated permitting and administrative procedures. They will also benefit from easier access to finance and adhere to public procurement measures designed to enhance the resilience and environmental sustainability of supply chains (incentives for locating these projects in areas more affected by the energy transition have also been provided). In general, to increase supply chains' resilience, the NZIA introduced new public procurement measures (e.g., for auctions and contracts for renewable energy projects) that disadvantage bids that involve the use of products from a third country holding more than 65% of the EU's market share in the sector concerned (another rule that would impact Chinese equipment). Moreover, for promoting and selecting Net-Zero Strategic Projects, a Net-Zero Europe Platform is established with the Commission, member states and industry. Yet, the NZIA did not provide for additional EU-level funding, as the major contribution to its objective is expected to come from the private sector. However, it allowed member states to use state aid for implementing strategic projects. In line with the logic of the IPCEI, this is another important derogation from the EU traditional view that prioritized market competition. For this latter purpose, the NZIA works in parallel to the Temporary Crisis and Transition Framework, which was issued in 2023 as an upgrading of a similar scheme launched in 2022 to address the energy crisis caused by the war and target low-carbon technology such as batteries, solar panels, wind turbines and electrolysers as well as the production and recycling of critical raw materials (Tagliapietra, Veugelers, and Zettelmeyer 2023). Under this framework, which is in place until 2025, member states can also match the amount of aid that a company could receive for an equivalent investment in other locations, a measure that was the most evident reaction to the US IRA (ibid.). Externally, the NZIA focused on industrial cooperation with like-minded countries through initiatives such as Net-Zero Industrial Partnerships or the Trade and Technology Council—like the one established in 2022 with India (Table 4.7)—as well as on working with the US to manage the consequences of the IRA (an US-EU Task Force on IRA was established on October 2022; see Chapter 5). With the NZIA, the Commission also relaunched the proposal for a common export credit strategy—including the idea of creating an EU export credit facility, already encompassed in the 2021 Trade Policy Review—in close coordination with EU financial tools in order to support the internationalization of European green business while enhancing the energy transition abroad (Table 4.8). A similar innovation, like the EU Sovereignty Fund, would push the bloc's green power strategy into the reign of economic power. However, at the end of 2023, it remained only at the discussion stage. Finally, the NZIA provided for the launch of the European Hydrogen Bank: a financial mechanism to support the uptake of green hydrogen production in the EU, but also (possibly) to promote imports in coordination with the EU Energy Platform.

For the CRMA, the starting point was the recognition that the EU would never be self-sufficient in supplying many minerals and materials critical for the green (and digital) transition; it will continue to rely on imports for the majority of its consumption, but this should be reduced and more diversified (European Commission 2023b). Like the NZIA, hence, the CRMA comprised internal measures and domestic targets for a list of critical raw materials, as well as a foreign policy dimension (Table 4.8). The targets (by 2030) pertain to the quota of extraction (10%), processing (40%), and recycling (25%) of critical minerals and materials (e.g., cobalt, copper, lithium, nickel, REEs, and graphite products) that should take place in the EU as well as a threshold (65%) for the maximum annual consumption from any stage of production and processing that can come from a single country. Like the NZIA, the CRMA did not foresee new EU funding. Rather, it has provisions on supply chain monitoring, stockpiling, coordination among member states, and the enhancement of recyclability. Moreover, to facilitate capacity expansion within the EU, it provides for enhancing and simplifying the regulatory framework for permits and access to finance to relevant Strategic Projects. Internationally, the CRMA focuses on further promoting bilateral strategic partnerships, and using trade deals as well as the Global Gateway investment agenda to improve the diversification of the EU supply chains (Strategic Projects under the CRMA can also be located in third states). In addition, the CRMA provides for the establishment of a Joint Purchasing mechanism to aggregate demand at the European level. Finally, although the EU had already joined the US-led Mineral Security Partnership—launched in 2022 and also comprising Australia, Canada, Finland, France, Germany, Japan, South Korea, Sweden, and the UK (see Chapter 5)—the CRMA envisages the creation of a Critical Raw Materials Club with like-minded countries. At the same time, it suggested working in the context of the WTO to tackle unfair trade practices and supporting global EU standards for critical raw materials supply chains. This latter objective is aligned with the EU Strategy on Standardisation and has a strong focus on environmental, social, and governance criteria.

The post-EGD agenda was also embedded in the bloc's regional strategies for the EU's neighbourhood and beyond (Tables 4.9 and 4.10). Within these regional strategies—which also aim at supporting the implementation of the Paris Agreement and third country's NDCs as well as promoting other traditional (and wider) EU goals, such as human rights, democracy, and good governance—the EU has sought to better integrate its green foreign energy policy tools: diplomatic, regulatory, and economic. These latter centre around the IPA-WBIF (for the Western Balkans) and the geographic envelopes of the EFSD+, which constitutes the larger part of NDICI-Global Europe allocations (around €60 billion out of the total €80 billion for the 2021–2017 period). The EFSD+ has not only consolidated blending and (expanded) guarantees operations in third states under a common governance framework but has also allowed for a larger participation of several financial institutions under the so-called open architecture. However, it has maintained a central role for the EIB as the major implementing partner in these operations (for instance, 67% of the EU's coverage provided by the EFSD+ Guarantee is dedicated to the EIB). At the same time, the EIB has been mandated to play a prominent role in the Global Gateway Strategy, which is implemented according to the Team Europe approach, i.e., via closer cooperation among EU institutions, member states and EU and national financial institutions. The EIB itself, in 2022, has created the EIB Global as its arm to support the Global Gateway Strategy and promote projects focused on climate action, renewable energy, transport, and sustainable development outside the EU. In 2022 alone, the EIB committed €1.6 billion for renewable energy projects abroad and another €0.3 billion for electricity networks (EIB 2023).

Overall, similar to previous financial tools (e.g., the regional blending facilities and the EFSD), many of these engagements still adhere to the conceptualization of green catalytic power. However, both the NDICI-Global Europe, with its policy first approach, and the Global Gateway Strategy place a stronger emphasis on promoting broader policy objectives via the investment projects they support. This includes initiatives aimed at enhancing the regulatory framework in third countries and exporting the EU's rules and perspectives on the energy transition. This (again) points to a closer integration among the various components of the EU's green foreign energy policy.

In the Western Balkans, the external influence of the EU green regulatory state has continued to expand (Table 4.9). This expansion is notably evident in the framework of the EnC, which encompasses some countries in the Eastern Neighbourhood as well. These countries have adopted key EU regulations, including the EU Governance Regulation (in 2021), the RED II directive, and other measures and climate targets outlined in the 2030 Climate and Energy Framework and Clean Energy Package (in 2022). In 2020, the EU initiated a Just Transition Mechanism for the Western Balkans to support these countries in transitioning away from coal (this mechanism includes Ukraine). The subsequent steps involve the export of the EU Climate Law and the ETS. These endeavours are complemented by sustained efforts to increase investment in renewables and electricity grids. Notable examples include the completion of the Trans-Balkan Electricity Corridor, rebranded under the new Global Gateway–Team Europe framework (Table 4.9).

The EU's engagement in the Eastern and Southern Neighbourhoods is characterized by a dual focus on shaping the policy environment and promoting investments. This approach, intended to benefit from market-oriented rules supported by the EU abroad, especially in terms of private investments, is evident in the ENP framework. This framework, encompassing both bilateral and multilateral aspects, ranks as the second largest geographical area in terms of EU financial commitments after sub-Saharan Africa (Table 4.9). Unlike the Eastern countries that joined the EnC, the EU's influence in these regions does not rely heavily on regulatory tools or strong conditionality to export its preferred policies. Instead, the EU utilizes softer channels, such as diplomacy and capacity building, to promote its green

Table 4.9 Greening the EU regional strategies for Western Balkans and the EU's neighbourhood

Region	Main EU initiatives/plans (*in italics*) and their key goals	Financial tools/commitment (*italics*)	Global Gateway–Team Europe (flagship energy transition projects)
Western Balkans	*Economic and Investment Plan for the Western Balkans—Green Agenda for the Western Balkans (2020)*: • Alignment with the EU Climate Law • Assistance for a possible expansion of the EU ETS • Western Balkans Just Transition Mechanism • Alignment with the acquis related to decarbonization of the energy sector in the framework of the Energy Community Treaty • Scale-up investment in renewable energy projects and energy infrastructure	IPA: *€5.8 billion (*)* WBIF: main vehicle for implementation (blending and guarantees)	• Trans-Balkan Electricity Corridor
Eastern Neighbourhood (ENP–Eastern Partnership)	*Eastern Partnership Policy Beyond 2020 (2020)—Economic and Investment Plan for the Eastern Partnership (2021)*: • Accelerate green recovery, growth, and climate action • Strengthen national energy legislative and regulatory framework in line with the Energy Community Treaty (where applicable) • Supporting renewable energy, cross-border energy connectivity and green hydrogen with energy sector reforms, capacity building and financial assistance	EFSD+ financial envelope for geographic programmes (*€19.3 billion for the Neighbourhood*)	• Georgia: Black Sea Connectivity (high-voltage submarine electricity cable to link Georgia, Azerbaijan, and Romania)
Southern Neighbourhood (ENP–UfM)	*A new Agenda for the Mediterranean (2021)—Economic and Investment Plan for the Southern Neighbourhood (2021)*: • Support green growth and climate action • Reinforced bilateral engagement and targeted assistance (regulatory and financial) to support large-scale investment in renewable energy and hydrogen production for both domestic consumption and export • Integration of regional electricity markets and networks • Support carbon pricing initiatives and the greening of the financial system (e.g., green taxonomies, green bonds)		• Tunisia: high-voltage electricity cable interconnection between Italy and Tunisia (ELMED interconnection). • Euro–Asia Interconnector • Morocco: green hydrogen project.

Note: (*) = IPA funds allocated for 2021–2027 to the green objectives of the Economic and Investment Plan for the Western Balkans
Source: Author's elaboration from EU documents and websites

agenda. These efforts encompass sectoral reforms, green taxonomies, and initiatives related to renewable hydrogen and electricity networks. Simultaneously, the EU engages in bilateral collaborations with specific countries, such as Morocco and Egypt. In terms of investment projects, the EU deploys its financial tools in a catalytic power fashion. Notable examples include ongoing support for projects like ELMED and the Euro-Asia Interconnector. The Black Sea Connectivity stands out as a high-voltage subsea electricity cable aiming to link the EU grid with partners in the South Caucasus. This initiative aims to support the energy transition in the region and reduce dependency on Russian fossil fuels (Table 4.9). Additionally, projects focusing on green hydrogen in Morocco underscore the upgraded energy cooperation between this country and the EU, marked by the signing of a Green Partnership in 2022 (Tables 4.7 and 4.9).

The EU's approach to regions beyond its immediate neighbourhood involves a combination of soft and catalytic power resources (Table 4.10). This perspective is evident in its engagements with various regions. Sub-Saharan Africa receives significant budget allocation under the EFSD+. In 2022, the EU launched the EU–Africa Global Gateway Investment Package, aiming to mobilize up to €150 billion under the Team Europe approach. Through this approach, the EU also supports the Just Energy Transition Partnerships involving South Africa and Senegal. Moreover, along with investments in renewable energy and electricity infrastructures, the EU has also focused on developing partnerships for critical raw materials supply chains and green hydrogen (Table 4.10). These sectors have also been included in the EU's engagements with Latin America and the Caribbean (Table 4.10). In the area of critical minerals, the EU has targeted especially resources-rich Argentina and Chile. Moreover, as this region has more developed green sectors and polices, the EU has also pushed for promoting its regulatory preferences and standards and for industrial goals, as in the case of EVs and batteries in Mexico (Table 4.10). In 2023, the EU also launched a Global Gateway Investment Agenda–Team Europe initiative for the region, targeting up to €45 billion in investments with a strong focus on green and climate action. In the Indo-Pacific, the last region in terms of the bloc's financial commitments, the EU has leveraged the EU–ASEAN relations for promoting cooperation on renewables, green industry, and hydrogen. In the cases of Indonesia and Vietnam, the Global Gateway–Team Europe approach has been deployed to support Just Energy Transition Partnership agreements. Funding commitments are made for the implementation of bilateral partnerships related to raw materials, batteries and hydrogen in Kazakhstan, a resource-rich country (see Table 4.7).

Table 4.10 Greening the EU regional strategies beyond the EU's neighbourhood

Region	Main EU initiatives/plans (*in italics*) and their key goals	Financial tools/commitment (*italics*)	Global Gateway–Team Europe (flagship energy transition projects)
Sub-Sahara Africa	*EU–African Union Joint Vision for 2030 (2022)—EU–Africa: Global Gateway Investment Package (2022), Africa-EU Green Energy Initiative (2022):* • Support investments in renewable energy, hydrogen, energy infrastructures as well as market integration and reforms (add 300 GW of renewable energy capacity by 2030) • Promote green bonds markets, and provide technical assistance to attract private investors • Support industrialization and promote investment for sustainable raw materials supply chains	EFSD+ financial envelope for geographic programmes (€29.2 *billion*) *2022: Global Gateway Investment Package-Team Europe: aims to mobilize €150 billion*	• Namibia: 85 MW solar PV and electrolysers (implementation of partnership on raw materials value chains and hydrogen) • Madagascar: 320 MW increase of renewable energy production • Benin, Côte d'Ivoire, Niger: solar PV plants and upgrading of electricity networks • Nigeria, Niger, Benin, Burkina Faso: high-voltage power interconnection line • Senegal and South Africa: investments via the Just Energy Transition Partnerships
Latin American and Caribbean	*A New Agenda for Relations between the EU and Latin America and the Caribbean (2023)—Third EU–CELAC Summit (2023)—The European Union—Latin America and the Caribbean Global Gateway Investment Agenda (GGIA) (2023):* • Increase regional and bilateral cooperation to accelerate a fair green transition • Partnership and trade agreements for sustainable critical raw materials supply chains with a view to promoting LAC countries' economic development • Promote investments and cooperation on regulatory framework for renewable energy, energy infrastructures, and green hydrogen • Promoting hydrogen markets based on common standards and certification schemes, including through the European Hydrogen Bank • Support taxonomies and green bonds initiatives	EFSD+ financial envelope for geographic programme (€3.4 *billion*) *2023: Global Gateway Investment Agenda-Team Europe: aims at mobilizing €45 billion*	• Argentina, Costa Rica, Colombia, Chile, Panama, Paraguay, Uruguay: production of green hydrogen • Argentina, Chile: development of critical raw materials supply chains for lithium and copper • Mexico: industrial parks, renewable energy, supply chains development for solar plants, EVs, and battery production • Argentina, Bolivia, Brazil, Colombia, Ecuador, Panama: renewable energy projects • Ecuador–Peru: power interconnection

Continued

Table 4.10 *Continued*

Region	Main EU initiatives/plans (*in italics*) and their key goals	Financial tools/commitment (*italics*)	Global Gateway–Team Europe (flagship energy transition projects)
Indo-Pacific	*EU Strategy for Cooperation in the Indo-Pacific 2021 EU–ASEAN Energy Dialogue 2022*: • Mobilize energy dialogues, partnerships, and financial instruments for sustainable, secure, and affordable energy • Prioritizing a just transition towards a decarbonized energy system • EU–ASEAN Energy Dialogue: policy dialogue, and cooperation programmes on renewable energy and hydrogen	EFSD+ financial envelope for geographic programme (*€8.5 billion*)	• Indonesia and Vietnam: investments via the Just Energy Transition Partnerships • Kazakhstan: implementation of the partnership on raw materials, batteries, and renewable hydrogen • Mongolia: investments in upgrading electricity networks

Source: Author's elaboration from EU documents and websites.

Finally, in 2023, the CBAM became operational. Its primary goal is to address the risk of production moving outside the EU to countries with less stringent climate polices and/or lower carbon prices. This potential movement could undermine both European industry and climate policy objectives. The CBAM runs in parallel with the EU ETS and is initially applied to specific imported goods covered by this system, including cement, iron and steel, fertilizer, electricity, and hydrogen (Wettestad 2023). Although this is a unilateral regulatory measure, which could possibly expand the global reach of the EU's green regulatory power, the Commission has made efforts to ensure that its design is compatible with WTO rules. Moreover, in its transitional period, which will last three years, the CBMA requires only that importers of the above-mentioned goods report the GHG emissions embedded in their imports. The formal implementation of CBAM is set to begin at the start of 2026. Importers will be obligated to purchase CBAM certificates, with their prices linked to the EU ETS. Alternatively, importers can prove that a carbon price has already been paid during the production stage, allowing for the deduction of the corresponding amount.

5

The United States

A Reluctant Follower or a Re-emerging Leader?

Politicization, climate change, and renewables in the US

The history of (modern) renewable energy began in the US, where, in the 1950s, (silicon) photovoltaic technology for solar power was developed. In the 1960s, this innovation was perfected for the space and satellite industry by NASA, which also developed the first exemplars of lithium-ion batteries. In that period, thanks to the Mountain Pass deposit in California, the US was also a leader in REEs production, technological applications, and trade. In the 1970s, research on renewable energy, both solar (in 1974 it was created the Solar Energy Research Institute, SERI) and wind technology, increased alongside demonstration projects and commercialization. The late 1970s were, then, an important period for the development of the US renewable energy policy. In the wake of the oil shocks, the Carter administration (1977–1981), backed by a bipartisan consensus, increased federal funding for R&D on renewables, which were framed as promising technologies for reducing oil dependency and increasing national (energy) security. In 1978, Congress enacted the Public Utility Regularity Policy Act (PURPA), which initiated the deregulation and liberalization of the US power sector and provided the first incentives to produce electricity from renewables. The Energy Security Act of 1980 provided additional funding for research on renewables and biofuels.

In sum, after the oil shocks, the US was a green leader, and although the costs of wind and solar technology were still high, both were gradually being deployed throughout the power sector, where hydropower was already responsible for around 10% of electricity production in the early 1980s. To be sure, support for the fossil fuel industry and nuclear energy remained the main priority in Washington. Coal was still the backbone of the US power mix contributing to around half of the total electricity production. The Carter Doctrine had further oriented the US foreign policy towards prioritizing security oil supply from the Middle East. However,

Green Superpowers. Andrea Prontera, Oxford University Press. © Andrea Prontera (2024).
DOI: 10.1093/9780191987304.003.0005

the institutional innovations introduced in the late 1970s laid the foundations for the structuration of the US green developmental network state (e.g., Keller and Block 2011; MacNeil and Paterson 2012; MacNeil 2013, 2014). This emerging apparatus was organized around key federal agencies, chiefly the Department of Energy (DoE). The DoE was established in 1977 and, through its Office of Energy Efficiency and Renewable Energy, promoted the creation of several national laboratories, research institutes, and programmes, as well as demonstration centres that functioned in collaboration with universities and the industry. This system allowed the US to maintain a global lead in novel renewable energy technology. In turn, the technological advances originating from the federal R&D initiatives have been instrumental in reducing the costs of solar PV and wind turbines, favouring their commercialization by the nascent US green industry at home and abroad (Nahm 2021).

A developmental network state emerged in the US during the Cold War. This began with the creation, in 1958, of the Pentagon's famous Defence Advanced Research Projects Agency (DARPA) in the aftermath of the Soviet success with Sputnik 1. Unlike in Europe or China, the military sector and the Department of Defence (DoD)—with its large budget, expertise, bipartisan political support, and links with laboratories and industry—has traditionally fulfilled a vital role in the US developmental approach (e.g., Roland and Shiman 2002; Block 2008; Keller and Block 2011; Weiss 2014; Keller and Block 2015). This close collaboration, coupled with deep-pocket venture capital and various federally funded programmes, was able to sustain start-ups and innovative companies, allowing them to take off and compete on a global scale. To coexist with the peculiarities of the US political system and market ideology, however, the institutional structures of this apparatus have remained 'hidden' to the public (Block 2008). Moreover, the main focus of this developmental network state was not originally on renewable energy but on those technologies (e.g., biotechnology, information technology, software) that could more easily diffuse in the market. In any case, in response to both internal and external pressures, this institutional structure eventually oriented more specifically towards green technologies.

During the Reagan presidency (1981–1988), in the wake of the counter-oil shock and the rise of neoliberalism, attention to renewable energy decreased and several measures of the previous administration were halted. Federal funding for renewables drastically dropped, negatively affecting the US green industry (Moe 2015). Fossil fuel interests continued to shape and drive Washington's energy policy at home and abroad, and only nuclear energy increased its role in the country's electricity mix, despite an initial slowdown following the 1979 Three Mile Island accident.

In the wake of the first Gulf War, the administration of George H. W. Bush (1989–1993) issued the Energy Policy Act (1992), which was driven by traditional energy security concerns. This act further promoted the liberalization and deregulation of the power sector. However, it also relaunched R&D in the area of renewable energy—in 1991, SERI was upgraded into the National Renewable Energy Laboratory (NREL)—and established supporting measures for the deployment of wind and solar installations, particularly production tax credits and investment tax credits (tax incentives were also provided for EVs). During the 1990s (and the 2000s), however, supporting measures for renewables were renewed in an inconstant manner that was subject to congressional tug of war and opposition by fossil fuel interests and Republicans (Joskow 2001; Laird and Stefes 2009; Moe 2015; Stokes and Breetz 2018; Downie 2019). This period saw the birth of an important trend: while the institutions and policies of the US (green) developmental network state continued to expand—although not always at the same pace and still 'hidden' in order to cohabit with the US-style market ideology (Block 2008)—they were not accompanied by consistent nationwide measures to promote the diffusion of renewables, such as the feed-in tariffs or other types of subsides implemented in Europe and China, nor by measures to limit emissions, such as a cap-and-trade system. This trend was partially the result of the growing politicization of climate issues and the polarization of the public's views on the matter, which were paralleled with growing Republicans' opposition to stricter environmental regulations, including those targeting the fossil fuel industry or usage (Dunlap and McCright 2010; McCright and Dunlap 2011). These stressors, coupled with an aversion to binding international commitments that could undermine the national economy and limit sovereignty, also guided US external action. The George H. W. Bush administration successfully opposed the inclusion of legally binding targets and timetables for reducing GHG emissions in multilateral negotiations for the UNFCCC, calling instead for the formulation of more flexible national strategies (Bodansky 2001; McGee 2014). This more conservative position represented the early manifestation of the US green foreign energy policy. This was reflected in the final text of the UNFCCC, which was approved by the US Senate in 1992.

During the Clinton years (1993–2001), the US green developmental network state was expanded, and the administration framed renewables as a means of simultaneously reducing both dependence on foreign oil and GHG emissions (Simpson 2001; MacNeil 2014). Yet the supporting measures for the deployment of renewables continued to encounter problems due to the opposition in Congress. Hence, while the US maintained an edge in low-carbon technology innovation, its renewable energy policy remained poorly institutionalized, which had negative effects on the US green industrial sector. Abroad, the more conservative US position on global climate policy was maintained, although the Clinton administration adopted a more engaged attitude on the matter (Depledge 2005; Meckling and Van de Graaf 2020). In particular, it was able to include provisions in the Kyoto Protocol for market-based flexibility mechanisms, such as international emission trading, and clean development mechanisms (CDM). These mechanisms reflected Washington's preference for market-based forms of climate governance and, as illustrated in Chapter 3, were also used to provide support to renewable energy projects in developing countries like China. Conversely, the US was not able to overcome resistance from the G-77 plus China grouping, which refused to adopt emission reduction commitments. The final agreement encountered bipartisan opposition in the Senate—per the US Constitution, the Senate plays a crucial role in the approval of international treaties, which can pass only if supported by a two-thirds supermajority—which halted Clinton's international climate efforts. In the end, his administration was not able to ratify the Kyoto Protocol.

During the George W. Bush years (2001–9), US foreign energy policy, spurred by the military intervention in Iraq, focused on fossil fuels and supply security concerns. Moreover, the US withdrew from all further multilateral negotiations under the Kyoto Protocol (Depledge 2005). However, through the Department of State (DoS), Washington launched an alternative strategy based on the establishment of low-carbon technology (minilateral) initiatives—such as the 2005 Asia-Pacific Partnership on Clean Development and Climate and the 2007 Major Economies Meeting—and bilateral climate partnerships (e.g., McGee 2014; Meckling and Van de Graaf 2020). The latter were signed with several industrialized and developing countries, including Australia, Brazil, Canada, China, the EU, Germany, India, Italy, Japan, Mexico, New Zealand, South Korea, Russia, and South Africa (US Department of State 2008). This move represented a shift in US green foreign energy policy away from the UN multilateral framework and a sole focus on the climate issue area. The US-led minilateral initiatives and bilateral partnerships did not provide binding emission targets or compliance mechanisms. Instead, they emphasized cooperation on climate science and renewable energy technology, combining diplomacy, capacity-building, and soft-power efforts to co-opt, on a voluntary basis, other partners. This approach reflected that followed at home by George W. Bush's administration—the green component of the US developmental network state continued to expand in this period (McGee 2014)—an approach that considered technological advances in the area of renewables to be a better way to address climate change than enactment of stricter environmental regulations or emissions targets. In 2003, cooperation with China was institutionalized with the launch of the US–China Working Group on Climate Change. Moreover, the DoE and the Department of Transportation launched the International Partnership for the Hydrogen Economy to foster dialogue on hydrogen and fuel cell R&D and standards (this initiative was renamed, in 2009, the International Partnership for Hydrogen and Fuel Cells in the Economy).

Along with these initiatives, the George W. Bush administration deployed diplomacy and soft power resources to promote regulatory and market reforms in third states, especially in the power sector. Conversely, Washington did not deploy regulatory tools to favour the promotion of renewables abroad or to foster the adoption of a global climate regime in which it did not take part. After the end of the Cold War, the US pushed to expand its international liberal agenda, both through regional deals, like the NAFTA, and through the WTO. In the 2000s, the George W. Bush administration also negotiated several bilateral free trade agreements that included provisions for the promotion of core environmental protections and for supporting certain multilateral environmental agreements. Again, however, this 'regulation through trade' logic was not embraced by Washington in the area of climate change and mitigation policies.

The liberalization provisions of the NAFTA included coverage of renewable energy and renewables technologies trade as well as electricity trade. However, the formers were not the focus of this regime, which was also almost silent on climate change. The development of electricity interconnections and trade among the US, Canada, and Mexico, on the other hand, had begun well before the establishment of the NAFTA (grid interconnections between the US and Canada and the US and Mexico date back to the beginning of the twentieth century). In 2001, however, the US, Canada, and Mexico established the North American Energy Working Group to foster cooperation and energy trade. Electricity exports from Canada to the US (especially hydropower) have markedly increased since the mid-2000s, paralleled with closer regulatory cooperation and market integration (Parfomak et al. 2017). There are currently more than thirty power transmission linkages between Canada and some North American states. On a national basis, imports from Canada are quite small—they were less than 1.5% of total US electric power generation in the 2010s—but the integration of the US and Canadian networks improved grid reliability in both countries. Conversely, both infrastructural connections and electricity trade have remained very limited with Mexico (ibid.).

Under George W. Bush, domestic US renewable energy policy continued to develop in an unpredictable way. This negatively affected the deployment of wind and solar installations: in 2005 their contribution to electricity production was less than 0.5%, whereas coal still supplied almost 50% (Table 5.1). In the wake of the 2003 war in Iraq, the emphasis on improving US energy security was situated on traditional sectors backed by large industrial interests; i.e., on extracting more oil and gas at home—including shale oil and gas—and on developing nuclear energy, which like fossil fuels was supported by large subsides (Sovacool 2008). In this period, several US states enacted measures to support the deployment of renewables, mainly Renewables Portfolio Standards, which typically prescribe that a certain share of total annual electricity sales must come from renewables. This policy, initiated in 1997, rapidly diffused across the country; by the early 2010s, thirty states had established Renewables Portfolio Standards (NCSL 2021). The election of Barack Obama (2009–2016) signalled an important turn in internal and international US renewable energy policy. However, the peculiarities of the US political system, in which the presidential agenda can be halted in Congress by opposing political majorities, continued to hamper US green efforts at home and abroad.

The Obama administration embraced a pro-renewables and pro-climate action position. In the wake of the 2008 financial crisis, not only did it widen the green footprint of the US development network state, but it also adopted a more vertical (and visible) industrial policy approach to upgrade manufacturing capacity for renewables and low-carbon technologies, such as solar PV, wind, and EVs (e.g., Keller and Block 2015; MacNeil 2016; Block, Keller, and Negoita 2023). Indeed, the crisis shifted the state–market balance towards greater government interventionism in Washington, well beyond R&D, invention, and commercialization. This manifested in the American Recovery and Reinvestment Act of 2009 (ARRA) and several other measures that targeted traditional and emerging industries (Di Tommaso and Schweitzer 2013). Still, due to the resistance in Congress, mostly from Republicans but also from Democrats from districts with strong fossil fuel interests, the Obama green agenda was often halted (e.g., Moe 2015; Downie 2019). As in previous periods, financial support for the deployment of renewables remained volatile, and President Obama was not able to realize all the components of his administration's ambitious climate plan. Nevertheless, the growth of renewables accelerated in these years: wind and solar energy installed capacity increased from around 25,000 MW in 2008 to more than

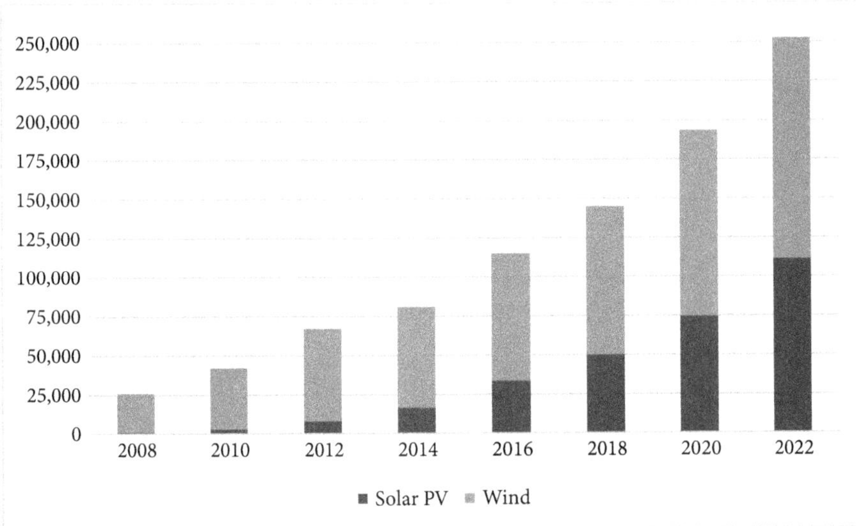

Figure 5.1 Solar and wind energy installed capacity in the US (in MW, selected years)
Source: Author's elaboration from IRENA (2018, 2023a)

100,000 MW in 2016 (Figure 5.1). In addition, under Obama renewables and low-carbon technology were increasingly reframed from a means for reducing energy dependency and addressing climate change to an opportunity to relaunch economic growth and jobs after the crisis, as well as a new field of global competition with major powers, notably China. The 2008 financial crisis momentum, in fact, intertwined with the (gradual) realization in Washington's policy circles that it was risky to shift from a dependency on Middle Eastern oil to a dependency on a potential rival in key low-carbon technology supply chains. In his 2011 State of the Union Address, President Obama explicitly called the period another 'Sputnik Moment', a critical juncture that, as did the Cold War space race, required active government support to ensure US global leadership in emerging green and high-tech industries (The White House 2011a).

This change was also the result of the shale revolution, which (ironically, under one of the more pro-renewables administrations that the country had had to that point) turned Washington from an importer to an exporter of oil and gas. To be sure, Obama's energy policy, although greener than that of his predecessors, still followed the so-called 'all-of-the-above strategy'; that is, it aimed to develop 'every source of American-made energy' (The White House 2014). During his two terms, US oil production swelled from around 8 million barrels per day in 2008 to around 15 million in 2016 (it reached

Table 5.1 Share of electricity production (percentage) by source in the US (2000–2022)

	2000	2005	2010	2015	2020	2022
Coal	51.7	49.8	44.9	33.1	19.1	19.2
Gas	16.1	18.8	24	32.6	40.1	39.3
Oil	3	3.1	1.1	1	0.8	0.9
Nuclear	19.8	19.3	19.6	19.5	19.5	17.9
Hydropower	7.1	6.5	6.1	5.9	6.9	5.9
Wind	0.1	0.4	2.3	4.6	8.3	10.1
Solar	0.01	0.01	0.03	0.9	3.2	4.7
Bioenergy	1.6	1.3	1.3	1.5	1.3	1.2

Source: Author's elaboration from Ritchie Roser and Rosado (2023)

20 million by the end of the decade; EIA 2023). The crude oil export ban in place since 1975 was lifted in 2015, and in 2020 oil exports surpassed imports. Natural gas production also increased significantly in the same period, and exports surpassed imports by 2017. This new bonanza favoured a shift from coal to natural gas for power generation (Table 5.1). However, it was the Trump administration that would leverage the fruits of the shale revolution for the US's international energy engagements. Under Obama, the US improved and expanded its green foreign energy policy beyond the climate and energy issue area, diplomatic tools, and soft power.

First, Obama re-engaged the US with the UN multilateral climate regime. Washington climate diplomacy was instrumental in brokering a deal that rescued failing negotiations at the 2009 COP 15 in Copenhagen (Christoff 2010). The US also endorsed the Green Climate Fund—established at the 2010 COP 16 in Cancun as the primary multilateral financing mechanism of the UNFCCC—and supported the introduction of the pledge-and-review model that set the stage for the 2015 Paris Agreement. This agreement, as it was not based on new international legal obligations, allowed Obama to avoid the involvement of Congress; the mandate provided by the UNFCCC, already ratified by the Senate, was sufficient for its approval. For the Paris Agreement, the Obama administration secured diplomatic cooperation, especially from China and India. In November 2014, Obama and Xi Jinping even jointly announced their 2030 climate targets: Washington pledged to cut its GHG emissions by 26–28% below 2005 levels by 2025.

Meanwhile, in 2011 the US joined the IRENA. The same year, the Bureau of Energy Resources (ENR) was created within the DoS, with the goal of strengthening US foreign energy engagements in both traditional fossil fuel sectors and in the area of renewables, electricity, and climate change. Moreover, the Obama administration promoted a set of additional (minilateral) initiatives for enhancing dialogue and cooperation on climate change, renewable energy, and low-carbon technology (Lesage, Van de Graaf, and Westphal 2010; Meckling and Van de Graaf 2020). Unlike under George W. Bush, however, these initiatives were intended to complement rather than challenge the UN climate regime. The Obama administration further institutionalized bilateral cooperation with China; it launched the US–China Clean Energy Research Center and several other platforms for fostering dialogue and policy exchange (see Chapter 3).

As mentioned, this cooperation with China was matched by growing concerns in Washington on Beijing's state-led green industrial rise. During the Obama presidency, the US took part in the emerging global trade disputes around renewable energy and green industrial policy. By the end of Obama's second term, the US Department of Commerce (DoC) and the US Trade Representative (USTR) had been involved in six out of sixteen such disputes, which addressed subsides and other non-tariff barriers, such as local content rules or price dumping, in the solar PV and wind industries (Meckling 2018). For example, in 2013 the US challenged, via the WTO, India's 'localization' rules for discriminating against US solar technologies (in 2016 the WTO ruled in favour of the US). Nevertheless, the most important case was opened in 2011 directly by the DoC against China in the area of solar PV. This dispute escalated over time, with the US government increasing trade tariffs and broadening their scope under Obama (2012, 2014) and Trump (2018; Beijing also responded by targeting US renewables support policies and local content rules). In addition, in the wake of the 2010 REEs crisis, the Obama administration increased attention to critical minerals and materials. In this area, Washington's domestic response was not particularly interventionist (Kalantzakos 2017a). However, in 2012 the US joined the EU and Japan in filing complaints in the WTO over Chinese REEs export restrictions. In 2014, the WTO ruled against China, and Beijing decided to comply with the provisions. However, this did not reduce China's prominent position within the global REEs supply chains, an issue that will remain on the agenda of subsequent presidents.

The focus on the trade and investment issue area characterized other important aspects of the US green foreign energy policy during the Obama presidency. In this period, the US engaged in multilateral negotiations under the WTO for an Environmental Goods Agreement, a process that, as already illustrated, was unsuccessful. Under Obama, the US also sought to give greater priority to climate change and renewables in the NAFTA agenda, in this case as well, with limited results. Obama, however, relaunched the country's global liberal agenda with the Trans-Pacific Partnership (TPP), finalized in 2016, and the Transatlantic Trade and Investment Partnership (TTIP). The latter provided for closer cooperation with the EU in the areas of sustainable development, renewable energy, and energy efficiency. The TPP (the members of which were Australia, Brunei, Canada, Chile, Japan, Malaysia, Mexico, New Zealand, Peru, Singapore, and Vietnam) was the centrepiece of Obama's strategic Pivot to Asia to counter China's expansion and limit the influence of its state-led model of capitalism (e.g., Leoni 2022). Indeed, this transregional deal was intended to both increase economic exchange and promote rules and standards that were in line with US views and interests, such as stricter protection of intellectual property, environmental obligations, commitments on worker rights, and provisions for limiting unfair competition by SOEs. The TPP also aimed to reduce barriers to the export of several renewable energy technologies in order to boost the competitiveness of the US green industry. Both initiatives would be halted by the Trump administration (the US withdrew from the TPP in 2017). However, under Obama, the US signed three bilateral free trade agreements with Colombia, Panama, and South Korea in 2011 (the latter of which facilitated a reduction in EV tariffs and increased related bilateral trade; each of these three agreements is still active).

Finally, under Obama, the US expanded the use of economic tools for its green foreign energy policy and green power strategy. In 2010 the Obama administration launched the Global Climate Change Initiative in order to scale up Washington's climate-related aid to developing countries, with a focus on promoting renewable energy (US Department of State 2016; Lawson 2017). These resources were channelled through multilateral programmes, such as the World Bank's Climate Investment Funds, which comprises the Strategic Climate Fund and the Clean Technology Fund; to the latter, the US contributed \$1.3 billion during Obama's presidency. In addition, \$365 million was granted to the Global Environment Facility. In 2014 Obama also pledged \$3 billion for the Green Climate Fund. However, because of opposition in Congress, his administration was ultimately able to grant only \$1 billion. Under Obama, the Department of the Treasury (USDT) formulated guidelines that made it more difficult for Western-led multilateral banks, such as the World Bank, to finance fossil fuels projects, ending support for new coal power plants in 2013. In 2015 the Obama administration committed the US to the new UN 2030 Agenda for Sustainable Development, which includes support for renewable energy.

For the international green agenda, Obama also mobilized the US development and foreign economic policy apparatus. The US Agency for International Development (USAID) had been involved in promoting (small) renewable energy projects for rural electrification in developing countries since the 1990s (e.g., Kozloff and Shobowale 1994). Under Obama, USAID work on renewables increased, and this was also the case for the Millennium Challenge Corporation (MCC). The MCC is an autonomous governmental agency, separate from the DoS and USAID, that was established in 2004 to channel US foreign aid (mainly grants) and encourage democratic governance and market reforms in low-income or lower-middle-income countries. In addition, the Obama administration mandated the US development financial institution—i.e., the Overseas Private Investment Corporation (OPIC)—and the US export credit agency—i.e., the Export–Import Bank of the United States (EXIM)—to promote solar and wind energy projects abroad and support the internationalization of national green business. The Overseas Private Investment Corporation was established in 1971 to mobilize private capital to address development challenges and advance US foreign policy objectives while helping US companies enter emerging markets. The Export–Import Bank of the United States became an independent agency, although still owned by the federal government, in 1945, with the goal of aiding the export of US goods and services (it operates following the OECD guidelines for export credit agencies). This deployment of Washington's economic tools highlights how, under Obama, the US green developmental state approach built up an external reach. However, the combination of development aid—often implemented in cooperation with several other private and public partners, including other development agencies—with the pro-national business mandates of institutions like the OPIC and EXIM has created a (partial) overlapping of green economic and catalytic power strategies. Indeed, as in the case of the EU, the US has also resorted to mechanisms of mobilization and connectivity in order to create multi-actor coalitions to better pursue its green foreign energy policy objectives. This catalytic power perspective is closer to the more archetypal liberal US understanding of state–market relations at home and abroad. Under Obama, such an approach was particularly evident with the Power Africa initiative, which was launched in 2013 as a flagship project of his administration to re-engage the continent and counter the Chinese BRI. This approach outlived his administration and continued into the following years.

In 2016, the election of Donald Trump resulted in a U-turn for US green action at home and abroad, reflecting the ongoing politicization and partisan division between Democrats and Republicans on the matter (e.g., Jotzo, Depledge, and Winkler 2018; Urpelainen and Van de Graaf 2018; Elkind 2020; Meckling and Van de Graaf 2020). Notably, as promised in the presidential campaign, in June 2017, Trump announced the US withdrawal from the Paris Agreement (a decision that was, however, reversed directly after Joe Biden's victory, which brought the US back into the Paris Agreement on January 2021).

Overall, under Trump, the US energy policy swung back to fossil fuels, with the goal of exploiting the results of the shale revolution, an approach that was reframed in terms of 'energy dominance' in the 2017 National Security Strategy (NSS 2017, 6). This perspective adapted to the new context some traditional elements that had characterized the US energy policy since the oil crisis and were closer to the Republicans' view on the matter. These elements comprise efforts to increase domestic production from traditional sources, remove regulations to make energy activities easier, and pursue (bilateral) energy trading opportunities with other countries (Ladislaw 2017). The Trump administration also dismantled Obama's main climate measures and weakened the US renewable energy policy. During the Trump years, solar and wind installations continued to grow, but at a slower pace than in the preceding period (Figure 5.1). The Trump administration attempted to dismantle the green turn instituted by Obama in the US developmental network state and industrial policy, as well as the economic tools for US green foreign policy that had been expanded during Obama's administration. In both cases, these efforts were contained by Congress: the 'renewable energy complex'—i.e., the ensemble of state–market relations, governance structures, and policies that have emerged around the US green sector—resisted the changes.

In fact, in the wake of growing rivalry with China, some elements of the US green foreign energy policy were further strengthened during the Trump presidency, such as measures in the area of critical minerals and REEs, the upgrading of the OPIC into the Development Finance Corporation, and the creation of the China and Transformational Exports Program, which was launched by the EXIM. Hence, although under Trump renewables were not prioritized in terms of contributing to climate action, jobs, or economic development, because of the need to keep pace with (and contain) China's green rise, low-carbon technologies became even more commonly perceived as strategic assets in the upcoming global competition with Beijing. Still, during the Trump presidency, the US did not formulate a well-defined green power strategy; actions in this field were a byproduct of the wider US global strategy of the period. The generally liberal agenda pursued by Obama shifted towards a more mercantilist approach that was based on a selective combination of economic rewards and punishments (e.g., Leoni 2022). This approach, which in some cases targeted even US allies (e.g., import tariffs on European steel and aluminium and on Canadian solar equipment), included halting the TPP and TTIP, renegotiation of the NAFTA (which was replaced by the USMCA in 2020), hindering the WTO (by disrupting the functioning of its Appellate Body), and direct confrontation with China through a trade war that began in 2018 and continued through the Covid-19 crisis and the end of the Trump presidency. This trade war involved tariffs and restrictions on several industries and technologies, including Chinese solar panels and EVs. However, again, this (bilateral) showcase of US economic might, in some cases accompanied by a 'de-coupling' rhetoric that was voiced directly by President Trump, was not part of a coherent approach to increase the US green global posture.

It was only with the election of Joe Biden that the US articulated a comprehensive foreign green energy policy and green power strategy. Between 2021 and 2024, the Biden administration resumed and enhanced the work initiated under Obama (Joe Biden was his vice president). In the wake of the Covid-19 crisis, it embraced an interventionist approach to relaunch the national economy, at the core of which were infrastructural investments, green manufacturing, renewable energy deployment, environmental justice, and climate action. In its first week, the Biden administration recommitted the US to the Paris Agreement, relaunched Washington's climate diplomacy—it established the Special Presidential Envoy for Climate (a position filled by John Kerry)—and enacted Executive Order 14008 on Tackling the Climate Crisis at Home and Abroad, which put climate action at the forefront of US domestic and foreign policy. In 2021 it created a National Climate Task Force, committed the US to achieving climate neutrality by 2050, and relaunched cooperation with allies, through the G7 and other initiatives, to promote its green agenda abroad. This agenda, which was initially formulated as part of the Build Back Better Plan, was mainly promoted at home through the 2021 Infrastructure Investment and Jobs Act (IIJA, also known as the Bipartisan Infrastructure Law), the 2022 CHIPS and Science Act, and most notably, the 2022 Inflation Reduction Act (IRA). Through these three pieces of legislation, the Biden administration further strengthened the US green development network state and green industrial policy, targeting key renewables and low-carbon technologies such as electricity grids, solar PV, onshore and offshore wind, EVs, batteries, and green hydrogen (the first US National Clean Hydrogen Strategy and Roadmap was issued in 2023).

Like Obama, Biden revitalized US green soft power and strengthened the economic tools of US green foreign energy policy. Unlike Obama's administration, however, Biden's did not try to promote an international liberal agenda. Under the heading of the 'new Washington consensus'—as coined by Jake Sullivan, Biden's National Security Advisor—the new US approach envisaged international agreements that were no longer focused on trade liberalization but on coordinating industrial policy measures and reinforcing supply chain resilience in strategic sectors (The White House 2023a). This approach, however, also comprised protectionist and mercantilist elements, as was well illustrated by the 'Buy American' provisions included in the IIJA and, even more so, by several IRA regulations that favoured national green industry—e.g., for the first time, the US has put in place local content requirements in clear violation of WTO rules (Kleimann et al. 2023, 3434)—and aimed to restructure low-carbon technology supply chains according to Washington's geopolitical preferences.

In addition, Biden adopted a more confrontational approach towards China than had been the norm in the Obama era, and he maintained the tariffs on several imports from China that had been introduced by Trump, including those on solar technology and EVs (during the Biden presidency, some Chinese PV companies were banned for human rights violations and forced labour under the 2021 Uyghur Forced Labor Protection Act). In the green realm, on the one hand, Biden resumed climate cooperation with Beijing; on the other hand, he markedly increased domestic and international efforts to reduce dependencies on China in critical low-carbon supply chains, simultaneously embracing a 'de-risking' rather than a 'de-coupling' narrative. This was the case even after the beginning of the war in Ukraine, which further highlighted the growing divide between the West and Beijing. In general, however, the measures taken by the Biden administration illustrate the emergence of a US green power strategy with a mercantilist edge, based on the deployment of the country's economic might and coupled with the military posture of a (declining) hegemonic power. This is exemplified by the inclusion of the US green turn as a core element of Washington's 2022 National Security Strategy (NSS 2022), as well as by the greening of the US network of alliances and strategic partnerships intended to counter China's rise, which stretched from the Indo-Pacific to the Middle East.

In the following section of this chapter, I illustrate in greater detail the evolution of the US renewable energy policy and green industrial policy. In particular, I highlight the innovations introduced in these areas under Obama, some of which survived under Trump and were relaunched and even expanded under Biden. I go on to explain the green foreign energy policy pursued by the US under the Obama and Trump administrations. Finally, in the last section, I analyse the changes in the US green foreign energy policy and green power strategy during the first years of the Biden presidency, when the links between internal and external green measures and between the US energy transition and its global strategy became more evident.

Another Sputnik moment?

As explained, the emergence of the first apparatus of what would become the US green developmental network state dates back to the Carter administration, which with the PURPA also enacted the first measures to support the deployment of renewables in the power sector (Table 5.2). During the Reagan administration, the main institutions and much of the green programmes created under Carter (e.g., the DoE, the DoE's national laboratory system, etc.) remained in place, although on a much smaller scale and with less visibility. However, no new measures for supporting the deployment of renewables were enacted through the 1980s. It was only with the Energy Policy Act of 1992 that research, investment, and deployment of different renewable energy technologies were relaunched (Table 5.2). With scarce political backing in Congress, however, the main measures supporting wind and solar installations—i.e., the investment tax credits (ITC) and the production tax credits (PTC) managed by the Department of the Treasury (USDT)—were renewed in an inconstant manner, generating uncertainty for investors and industries.

The US green developmental network state expanded during the Clinton and George W. Bush administrations. During these years, several federal programmes for renewable energy were launched, and R&D funding for wind and solar technology, EVs, batteries, hydrogen, and smart grids increased. The George W. Bush administration also enacted the Energy Policy Act of 2005, which (again) provided tax credits for wind and solar installations and established the first-ever tax credit for residential solar energy systems, i.e., the Residential Clean Energy Tax Credit (Table 5.2). Yet, also during the 2000s, the deployment of renewables continued to be hindered by limited political backing in Congress and the volatility of US renewable energy policy.

It was under Obama and his 'Sputnik moment' narrative that the US green efforts eventually escalated. His administration not only further upgraded the green dimension of the US developmental network state but, in addressing the 2008 economic crisis, embraced a more vertical and visible industrial policy that departed from the approach to innovation of the prior decades

Table 5.2 Main polices and supporting measures for renewable energy generation in the US (1978–2022)

Year	Policy/measure
1978	Public Utility Regulatory Policies Act (PURPA)
1992	Business Energy Investment Tax Credit (ITC) (relaunched with the Energy Policy Act of 1992, and renewed seven times until 2022)
1992	Renewable Electricity Production Tax Credit (PTC) (established with the Energy Policy Act of 1992, and renewed eight times until 2022)
2005	Residential Clean Energy Tax Credit (established with the Energy Policy Act of 2005, and renewed six times until 2022)
2009	American Recovery and Reinvestment Act (renewed and expanded PTC and ITC) (*)
2022	Inflation Reduction Act (renewed and expanded PTC and ITC) (**)

Note: (*) = the ARRA also established the 1603 Cash Grant programme (operative until 2012) that provided federal funding worth $26.2 billion for wind and solar projects. This resulted in 30,000 MW of new installed capacity (USTDA 2018). (**) = ITC and PTC, renewed with IRA, should expire in 2024, superseded respectively by the Clean Electricity Investment Credit and Clean Electricity Production Credit. According to the IRA, the phase out of these incentives will start the later of 2032 or when US greenhouse gas emissions from electricity are 25% of 2022 emissions or lower.
Source: Author's elaboration from Cunningham and Jordan (2023).

(e.g., MacNeil 2014, 2016). While the efforts of previous administrations had centred on R&D, on facilitative regulations, on broad industries, on coordinating various public–private collaborations, and on broad technologies and their commercialization, under Obama the federal government moved more decisively into the field of supporting specific technologies and firms, i.e., 'picking winners' and designating (private) national champions. These changes were complemented by ambitious climate policies, which were considered a top priority in the presidential agenda, as well as by the relaunch, mainly through the ARRA, of supporting measures for renewables deployment (Table 5.2). In addition, through the Department of the Interior (DOI), the Obama administration implemented a very generous policy for permitting renewable energy installations on federal lands.

With the ARRA, passed in 2009, the Obama presidency injected more than $800 billion into the US economy and provided approximately $90 billion in funding for several energy technologies (The White House 2016). The lion's share of resources targeted solar PV and wind energy, but significant funding was also allocated to the modernization of electricity grids, to EVs, to batteries, and to expanding industrial manufacturing capacity. This latter focus on manufacturing—for example, the ARRA provided $2.3 billion in tax credits for facility investments in renewable energy technologies—and on a set of specific sectors deemed to be of strategic significance, well illustrate the shift from the previous horizontal approach.

Along with the individual ARRA programmes, the Obama administration also increased federal R&D for renewables. The DOE launched a flotilla of major research centres and laboratories, such as the Energy Frontier Research Centers and the Energy Innovation Hubs (e.g., the Joint Center for Energy Storage Research, Critical Material Innovation Hub; MacNeil 2016; Block, Keller, and Negoita 2023). The administration also strengthened a new DOE agency, the Advanced Research Projects Agency-Energy (ARPA-E). This agency, which focused on breakthrough innovations and was modelled on the DARPA, was created in 2007 by the previous administration. It received an initial $400 million in funding from the ARRA and, later on, additional resources (Figure 5.3). Moreover, through the DoE, the Department of Commerce (DoC), and the DoD, the Obama administration established a network of Advanced Manufacturing Institutes as public–private partnerships funded jointly by the federal, state, and local governments, the industry, and the universities (the DoE alone launched six Clean Energy Manufacturing Innovation Institutes).

The new approach and green focus of the Obama administration were highlighted by the 2011 A Strategy for American Innovation: Securing Our Economic Growth and Prosperity and by its 2015 review (A Strategy for American Innovation). Both of these documents stressed that the administration's first priority in the field of innovation for fostering 'American leadership' was to 'unleash a clean energy revolution' based on solar and wind energy, advanced green manufacturing, EVs, batteries, and the modernization of electricity grids (The White House 2011b, 2015).

Another important illustration of Obama's more interventionist approach was the extended use of the DoE's Loan Guarantee Program, which deployed state capital through loans and guarantees to sustain a series of (private) firms and their selected technologies. Like the ARPA-E, it was originally established under the George W. Bush administration as part of the Energy Policy Act of 2005. Under Obama, mostly through the ARRA, the Loan Guarantee Program was expanded and directed (mainly) towards renewables and low-carbon technology. This led to the creation of the Advanced Technology Vehicle Manufacturing (ATVM) loan programme and so-called Section 1705 for large-scale solar and wind projects, solar and wind manufacturing, and EV manufacturing. Overall, during the Obama years, around $14.5 billion in loan guarantees were issued for wind and solar projects and to finance solar manufacturers, plus another $8 billion for the automotive industry, EVs, and batteries (US Government Accountability Office 2014). This included financial resources for Tesla but also more controversial funds for the failed PV company Solyndra. However, the initiative was largely successful, with only five of thirty-eight borrowers eventually defaulting (ibid.).

It is worth noting that the role public venture capital played in this resurgence was not novel within the US developmental state approach. Several federal agencies with a technology-focused mission have long used public venture capital to fund private companies and start-ups as a means of stimulating technological innovation and commercialization (Block and Keller 2011). Yet, under Obama and through the DoE, the federal government dramatically widened this perspective. Public venture capital is also important in the US developmental state approach because it works in conjunction with private venture capital. The latter often waits for DoE validation or support before investing in innovative firms (e.g., Block and Keller 2011; MacNeil 2014). Finally, in the wake of the 2010 REEs crisis, the Obama administration launched new initiatives on critical minerals. These, however, were less interventionist and focused on increasing R&D and establishing new bodies—such as the Critical Materials Institute created in 2013 by the DoE—rather than on upgrading industrial capacity.

In sum, the Obama administration strengthened the US green development network state and industrial policy, fostering a new approach and means of intervention (Figure 5.2). Importantly, these advances had an external reach, as under Obama the US sharpened the economic tools of its green foreign energy policy (see below). At the same time, the DoD increased its involvement in the field of renewable energy (e.g., Moe 2015; MacNeil 2016). The wars in Iraq and Afghanistan highlighted familiar problems related to reliance on oil and conventional fuels for military operations; coupled with the huge cost of fuelling the vast US military apparatus, since 2008 there has been a consistent push to increase the military's use of renewable energy. The DoD has established a series of programmes aimed at funding R&D and procurement of novel renewable technologies, including batteries, advanced EVs, and microgrids that could make military bases autonomous, at least as far as energy is concerned. Due to its very large budget, being sheltered from partisan battles, and its enormous size and energy demands, the military provides a huge market for renewables and low-carbon technology (e.g., Weiss 2014).

President Obama's ambitious green agenda encountered several problems that began in 2010, when the Republicans regained control of Congress. When the ARRA's funds were exhausted, the administration had limited

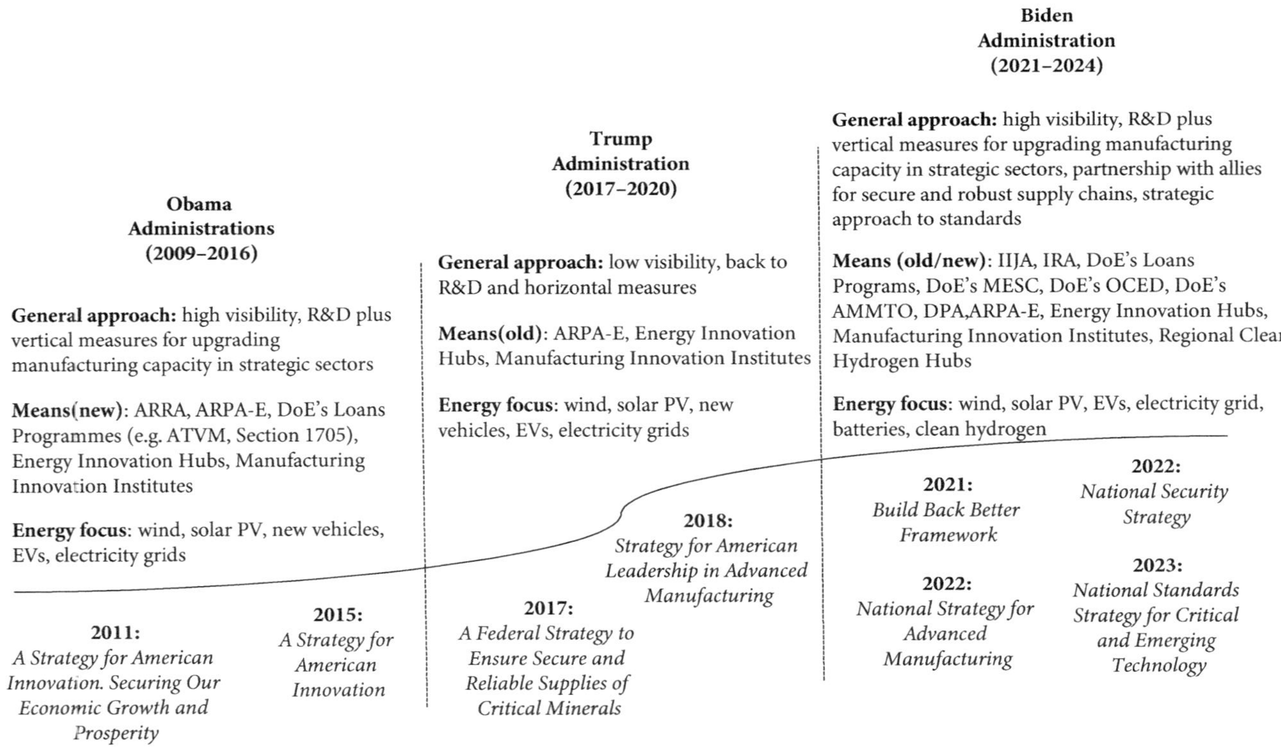

Figure 5.2 US green innovation and industrial policy under Obama, Trump, and Biden

Source: Author's elaboration. Italics indicate illustrative policy documents for each period

ability to renew financial support, especially for upgrading manufacturing capacity and supporting the deployment of renewable installations (for example, Section 1705 was discontinued in 2011). This affected the US green industry: some companies went bankrupt, and some moved to China or made deals with other foreign firms (Block, Keller, and Negoita 2023). Nevertheless, the mix of measures enacted under Obama resulted in a boost in wind and solar installed capacity, which allowed the US to surpass the planned targets (Table 5.3).

The climate component of the Obama administration's green agenda was more problematic. In 2009 the president made international commitments to reduce US GHG emissions to 17% below 2005 levels by 2020. Later on, he raised the target and, with the 2015 US NDC, pledged to reduce GHG emissions by 26–28% by 2025 (Table 5.3). In his first term, because of the opposition in Congress, Obama failed to pass nationwide 'cap-and-trade' legislation. Hence, in his second term, he intervened on the basis of presidential executive authority (Konisky and Woods 2016). In 2013 the Obama administration issued its Climate Action Plan and, through the Environmental Protection Agency (EPA), promulgated new regulations to increase energy efficiency standards, reduce emissions in the transport sector, and with the Clean Power Plan, reduce emissions in the power sector. This latter plan was

Table 5.3 Main climate and renewable energy targets and achievements in the US

	Obama 1st term	Obama 2nd term	Achieved (*)	Biden Administration
Climate	17% reduction in GHG emissions from 2005 levels by 2020	26–28% reduction in GHG emissions from 2005 levels by 2025 (NDC 2015)	2020: 17%	50–52% reduction in GHG emissions from 2005 levels by 2030 (NDC 2021) Climate neutrality by 2050
Renewable energy	Doubling renewable energy generation capacity from wind and solar by 2012	Doubling renewable energy generation capacity from wind and solar by 2020	2012: 67,212 MW (from 25,000 MW in 2008) 2020: 193,627 MW (from 80,875 in 2014)	Reaching 100% carbon pollution-free electricity by 2035

Notes: (*) = for GHG reduction, data from US Department of State and Executive Office of the President (2021); for renewable energy, see Figure 5.1
Sources: Author's elaboration

expected to accelerate the retirement of coal plants and inefficient gas-fired plants. However, its implementation was very difficult. More than twenty US states introduced legal challenges to the Clean Power Plan, and it was eventually halted by the Trump administration, which instructed the EPA to issue much weaker regulations for limiting CO_2 emissions in 2019 (the so-called Affordable Clean Energy Rule).

Notwithstanding these difficulties, the US was able to achieve its climate targets (Table 5.3). The government's green policies contributed to this goal, but a crucial contribution to the reduction in CO_2 emissions was the increase in shale gas production and the related decline in the price of natural gas, which favoured the shift from coal-based to gas-based power generation (Robinson 2013).

The Trump administration successfully dismantled Obama's climate policy. It also weakened the US renewable energy policy. His administration did not renew supporting measures for wind and solar energy, such as tax credits, and deprioritized renewable energy development on federal lands. However, it did not dismantle the US green developmental network state. Since its beginning, the Trump administration embraced traditional Republican rhetoric against 'big government'. It proposed several drastic budget cuts that targeted many of the initiatives of the previous administration, including the ARPA-E and other DoE green programmes (Elkind 2020). However, Congress—in 2018 the Democrats regained control of the House of Representatives—did not follow these requests and continued to fund such programmes at levels similar to, or even higher than, those of the previous years (Figure 5.3).

Overall, the US renewable energy complex was able to persist, despite the Trump administration's energy dominance approach centring on fossil fuels, CCUS technology, and nuclear energy rather than on solar, wind, or EVs. This green component of the US developmental approach and energy policy, however, lost its visibility in the presidential agenda, and the more vertical measures of industrial policy pursued under Obama, such as the DoE's loans programme, which had already been discontinued a few years before, were not relaunched. At the same time, as highlighted by the 2018 Strategy for American Leadership in Advanced Manufacturing, the Trump administration acknowledged the importance of sustaining manufacturing (see Figure 5.3). In 2020 it also launched two new advanced manufacturing institutes (The White House 2018). In addition, fuelled by President Trump's 'America First' rhetoric, it took a more confrontational approach vis-à-vis China in order to protect US innovations and industries. Yet, apart from the imposition of new tariffs on Chinese solar equipment and EVs in

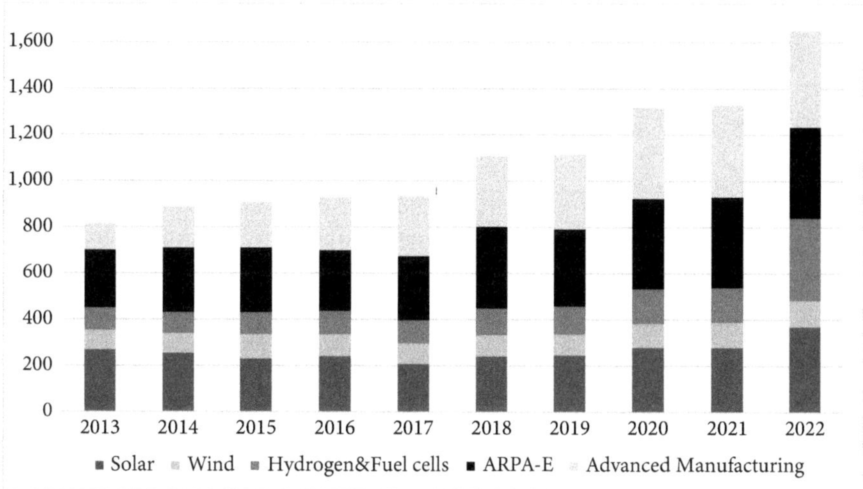

Figure 5.3 Annual funding (in million $) for selected DoE's R&D, demonstration, deployment, and manufacturing programmes (2013–2022)
Source: Author's elaboration from data provided in Cunningham and Jordan (2023). For 2022, the figure includes $80 and $200 million, respectively for solar and hydrogen&fuel cells, granted by IIJA. Data for advanced manufacturing is based on the funding for the Advanced Manufacturing Office manged by the DoE's Office of Energy Efficiency and Renewable Energy. In 2022, the Advanced Manufacturing Office was renamed as Advanced Materials and Manufacturing Technologies Office.

2018, the administration did not act directly to reduce US dependency in low-carbon supply chains. An important exception was the field of REEs and critical minerals, where again competition with China was a key driver, along with the wider national security implications of the matter.

The emphasis on these vulnerabilities was made clear before the Covid-19 crisis. In December 2017, Trump issued Executive Order 13817 (A Federal Strategy to Ensure Secure and Reliable Supplies of Critical Minerals), which tasked the DoC to formulate a comprehensive strategy and the Secretary of the Interior to draft a list of critical minerals. This list was issued in 2018 and included thirty-five minerals, including cobalt, lithium, and REEs. The strategy drafted by the DoC, which was realized in 2019, also had an external dimension (see below). It called for the acceleration of domestic critical resources development and the creation of more robust manufacturing capabilities (US Department of Commerce 2019). In 2020, under Executive Order 13817, the DoD signed new contracts with several REEs producers to strengthen the domestic supply chain, including through the use of the Defence Production Act (DPA; e.g., to finance the US company MP Materials, which owns California's Mountain Pass, the largest REEs mining operation outside China; US Department of Defence 2020). The same year, on 30 September, Trump issued another executive order—Executive Order 13953 on Addressing the Threat to the Domestic Supply Chain from Reliance on Critical Minerals from Foreign Adversaries and Supporting the Domestic Mining and Processing Industries—that adopted a more urgent tone and a sharper approach towards China. It was, however, the Biden administration that implemented a comprehensive strategy on low-carbon technologies and critical minerals supply chains linked to the needs of the US energy transition.

By rejoining the Paris Agreement, President Biden made new commitments to reducing GHG emissions 50–52% below 2005 levels by 2030, as highlighted in the country's 2021 NDC (Table 5.3). The Biden administration also pledged to accelerate the deployment of renewables to reach 100% carbon pollution–free electricity by 2035 and to achieve climate neutrality by 2050 (Table 5.3). The EPA, in 2021, issued stricter emission standards for vehicles and, in 2023, new regulations aimed at reducing emissions from coal and natural gas power plants, revising Trump's laxer provisions. Like Obama, Biden made it easier to access federal land (and waters) for renewable energy projects and relaunched supporting measures for the deployment of solar and wind installations.

Overall, Biden's administration redirected the US green innovation and industrial policy back on the path set by Obama (Figure 5.2). The 'Sputnik moment' window that had almost closed under Trump was fully reopened under Biden. Renewables, low-carbon technology, and the energy transition returned to the centre of the presidential agenda, framed (again) as a vehicle for fostering economic growth and jobs and enhancing US global technological leadership. This view was articulated in 2021 with the Build Back Better Plan. Several proposals from this plan, although often adjusted and downscaled, were then included in three crucial legislative acts approved by Congress between 2021 and 2022: the IIJA (signed into law by Biden in November 2021), the CHIPS and Science Act (signed into law by Biden in August 2022)—both passed with a large bipartisan majority—and the IRA (signed into law by Biden in August 2022), which was passed only by the Democrats. These acts also incorporated the new emphasis on critical supply chain diversification that had been spurred by the Covid-19 crisis and the growing rivalry with China and then further aggravated in the wake of the Russian invasion of Ukraine. This attention to critical supply chains had been evident since the beginning of Biden's presidency. On 24 February 2021, a few weeks after his inauguration, Biden issued Executive Order 14017 on America's Supply Chains, which mandated a comprehensive hundred-day review by the US government to identify risks, address vulnerabilities, and develop a strategy to promote the resilience of critical supply chains, specifically those related to semiconductor chips, pharmaceutical products, batteries (particularly lithium-ion batteries for EVs and stationary systems), and critical minerals and materials. The results of the hundred-day review—the DoE and the DoD were responsible, respectively, for the batteries and critical minerals and materials sections—recommended more interventionist measures to increase domestic manufacturing and industrial capacity (The White House 2021).

The IIJA provided large budgets for many activities already authorized by the Energy Act of 2020, which was signed by Trump after the 2020 presidential election but before the beginning of the new presidency. As part of its roughly $1 trillion commitment over ten years, the IIJA allocated more than $60 billion to the DoE for advancing R&D (see also Figure 5.3) and promoting green energy technology and manufacturing. To implement the IIJA, the DoE created, in 2021, the new multi-technology Office of Clean Energy Demonstrations (OCED), which was allocated a five-year budget of over $20 billion, of which $8 billion was earmarked for establishing (up to ten) Regional Clean Hydrogen Hubs across the country. Finally, the IIJA poured additional billions into the deployment of electricity grids ($75 billion), EV charging networks ($7.5 billion), EV battery manufacturing facilities, and the development of solar and wind technology (Ewing et al. 2022). The CHIPS and Science Act designated advanced energy technologies as one of ten fields to prioritize in order 'to keep the US the leader in the industries of tomorrow' (The White House 2022a). To this end, it authorized further budget increases for many DoE R&D programmes and enhanced the DoE's commercialization efforts.

It was with the IRA, however, that Biden took his most important step towards addressing climate change and speeding up the energy transition. To meet these goals, it provided nearly $400 billion in federal funding over ten years, primarily via subsidies and tax credits for consumption, production, and investment. It targeted several renewable energy and low-carbon technologies, especially solar and wind energy (onshore and offshore), EVs, batteries, hydrogen, and critical minerals (the IRA also targeted energy efficiency, nuclear energy, and CCUS; for details on the IRA climate- and energy-related measures, see Ewing et al. 2022; Ramseur 2023). These incentives aimed to accelerate the deployment of renewable energy (the IRA also invested in improving permitting procedures), increase the industrial capacity for low-carbon technology, and strengthen supply chain resilience. Several of the IRA's green provisions included protectionist elements, such as local content requirements prohibited under the WTO and regulations related to promoting the reorganization of supply chains towards North

America and Washington's trade partners and allies (other IRA environmental justice provisions created incentives for investing in specific regions of the country and for favouring vulnerable communities). For example, additional ('bonus') tax incentives are offered by the IRA for solar and wind (offshore and onshore) projects if 40% of the manufactured equipment is made in the US (55% from 2026). Moreover, of the $7,500 IRA tax credit for EVs, half is linked to the origin of battery components—50% should come from North America (US, Canada, or Mexico)—and the other half to the sourcing of their critical minerals. In this latter case, 40% of the value of the battery's critical minerals should be extracted or processed in the US (or recycled in North America) or in a country that has a free trade agreement with the US. In both cases, the thresholds are set to increase 10% each year, i.e., they will reach 80% in 2027. In order to strengthen the security of supply chains, the IRA also stipulated that, from 2024, EVs would not be eligible for tax credits if their battery contained components manufactured or assembled by a so-called 'foreign entity of concern' (FEOC)—possibly China, Russia, Iran, or North Korea—or, from 2025, if the battery contained critical minerals extracted, processed, or recycled by such entities. These latter measures, in particular, highlight how Washington has built up a green economic power strategy by leveraging its market and financial capabilities through targeted regulations and economic rewards. These measures have also triggered important repercussions for US green foreign energy policy (see below).

Finally, the IRA also allocated nearly $12 billion for the DoE's loan programmes, including $3 billion to relaunch the ATVM Loan Program and $8.6 billion for an upgraded Title 17 Clean Energy Financing Program centred on the deployment and manufacturing of solar, onshore and offshore wind, hydrogen, HVDC transmission lines, energy storage systems, EV charging infrastructure, electric grid components, and critical minerals processing, manufacturing, and recycling. In 2022, in order to implement the new provisions on manufacturing and supply chain expansion, the DoE created another new body, the Office of Manufacturing and Energy Supply Chains (MESC), and upgraded its Advanced Manufacturing Office into the Advanced Materials and Manufacturing Technologies Office (AMMTO). Moreover, in 2022 Biden authorized the DoE to use the DPA to further expand US manufacturing capacity for a set of critical renewable energy technologies, including PV modules and module components, electrolysers, and components for power grid infrastructures (The White House 2022b). In the same year, the DoD increased its commitments in the areas of renewables and low-carbon technology. It issued its first ever climate strategy, which set out ambitious goals, such as providing 100% carbon pollution–free electricity for army installations' needs by 2030, installing microgrids on every installation by 2035, achieving net zero emissions from army installations by 2045, and increasing the use of EVs in its vehicle fleet (US Department of Defence 2022). The DoD also increased coordination with the DoE to include in the US National Defence Stockpile those critical minerals necessary for the energy transition alongside those needed for defence purposes. In 2023, after years of defunding, $1 billion was allocated to increase the stockpiling of all these resources.

In sum, under Biden, the US green innovation and industrial policy reacquired high visibility and relaunched vertical measurers for upgrading manufacturing capacity in strategic sectors—combining old and new means—with the goal of improving supply chain diversification and resilience and better linking the internal and external dimensions of the US energy transition (Figure 5.2).

Along with renewables, as highlighted, the Biden administration increased the focus on clean hydrogen. The US produces approximately 10 million metric tonnes of hydrogen per year, mainly from natural gas, and it is second only to China in this field (Eicke 2023). The first US hydrogen strategy—the National Clean Hydrogen Strategy and Roadmap—was issued in 2023, setting a target of 10 million tonnes of clean hydrogen per year by 2030. The US renewables potential and freshwater resources are large enough to meet the domestic demand and (possibly) enable the production of green hydrogen for export in the long term. However, Washington's strategy is not centred only on green hydrogen, as it aims to support hydrogen production from diverse sources, including fossil fuels with CCUS. Moreover, although the National Clean Hydrogen Strategy and Roadmap mainly has an inward-looking perspective, prioritizing domestic production and infrastructures, it also includes an external dimension (Eicke 2023). This comprises promoting cooperation with key partners—in 2021 the US had already established bilateral partnerships with the Netherlands and Germany—on R&D and the harmonization of regulatory frameworks and standards through multilateral initiatives, such as those under the G7 and G20, and bodies like the Clean Energy Ministerial and the International Partnership for Hydrogen and Fuel Cells in the Economy.

This latter focus on standards is another novelty advanced under Biden. His administration issued, in May 2023, the National Standards Strategy for Critical and Emerging Technology, which targets those technologies that 'are essential for US competitiveness and national security' (The White House 2023b, 6). This strategy, which applies to renewable energy generation and storage systems, stressed that the long-standing leadership of the US in international standard-setting bodies was challenged by 'strategic competitors' such as China (ibid., 12). It reaffirmed the importance of maintaining the US private sector–led approach to standards development, but it also called for more coordinated and strategic action on the matter by the US government, including the promotion of standards through Washington's international partnerships and diplomatic engagements.

Energy dominance versus the renewable energy complex

The innovations introduced at home during the Obama era have been paralleled by changes in the US green foreign energy policy and green power strategy. Yet, under Obama, there were also some continuities with the previous period. Like his predecessor, Obama launched a set of minilateral forums—gathering major industrialized and emerging economies (including China and the EU)—to increase cooperation on renewable energy and low-carbon technology, such as the Major Economies Forum on Energy and Climate (2009), Clean Energy Ministerial (2010), and Mission Innovation (2015). This approach, however, was intended to complement US diplomatic engagement in the multilateral climate regime, and the Obama administration continued with the US's soft power efforts to support the global diffusion of regulatory and market reforms in the energy realm, notably with the Energy Utility Partnership Program for the power sector. Through the USAID (mainly) and the DoS's Bureau of Energy Resources (DoS-ENR), the administration also launched several other global, regional, and bilateral programmes to promote not only regulatory and market reforms but also investment in renewables in the Global South (Table 5.4; Table 5.5). These programmes—often developed in cooperation with NREL or other research institutes—combine capacity building, technical assistance, and limited financial support for project development, based mostly on grants, with the goal of favouring the diffusion of renewables by shaping the regulatory and policy framework of third states. In this indirect way, from 2010 to 2017, the USAID's energy programmes contributed to the installation of more than 120,000 MW of renewable capacity across twenty countries and facilitated more than $150 billion in private investment in new green generation capacity (USAID 2020).

Table 5.4 Main global and regional programmes for the power sector and renewable energy managed by the USAID and the Department of State (since 2007)

Geographical scope	Programme	Leading agencies (other agencies/ institutions involved)	Years	Objectives	Potential target regions/countries
Global	Energy Utility Partnership Program	USAID	2007–ongoing	Promote energy security, and renewable energy by providing capacity building to local utilities	Africa, Asia (excluding China) and the Middle East, Caribbean, Latin and Central America
	The Power Sector Programme	DoS–ENR	2009–ongoing	Promote regulatory reforms in the power sector to attract private investments (including for renewable energy)	
	Energy Regulatory Partnership Programme	USAID	2016–ongoing	Promote regulatory reforms, sharing best practices to help attract investments, and advance clean energy	
	Scaling Up Renewable Energy	USAID	2017–ongoing	Promotion of policies that enable renewable energy markets via private investment, and auctions for renewable energy installations	
	Strengthening Utilities and Promoting Energy Reform	USAID	2020–ongoing	Promote regulatory reforms for the electricity sector. Increase transparency, and support private investment	

Continued

Table 5.4 *Continued*

Geographical scope	Programme	Leading agencies (other agencies/ institutions involved)	Years	Objectives	Potential target regions/countries
Regional	Power Africa Initiative	USAID (DoS, DoC, DoE, USDA, USDT, USTDA, EXIM, OPIC/DFC, MCC) (*)	2013–ongoing	Doubling access to electricity in sub-Saharan Africa (with a target of 30,000 MW by 2030), promoting electricity networks, regulatory reforms and private investments	Sub-Sahara Africa
	Caribbean Energy Security Initiative	DoS–ENR	2014–2021	Promote energy security, renewable energy, regulatory reforms, and support private investment in the power sector	Jamaica, Eastern Caribbean
	Caribbean Clean Energy Programme	USAID	2015–2018	Promote renewable energy, regulatory reforms, and support private investment	Jamaica, Eastern Caribbean
	Scaling Off Grid Energy	USAID	2016–ongoing	Promote off-grid energy market to improve access to modern, clean and affordable electricity	Sub-Saharan Africa
	Clean Power Asia	USAID	2016–2021	Promote regulatory reforms, private investment, renewable energy, and regional market integration	ASEAN countries and Mekong region countries

Power the Future	USAID (DoS–ENR)	2017–2022	Accelerating the energy transition, diversifying energy sources, and creating competitive markets across Central Asia	Central Asia (Kazakhstan, Kyrgyz, Tajikistan, Turkmenistan, and Uzbekistan)
Asia EDGE (Enhancing Development and Growth through Energy)	DoS–ENR (USAID, USDT, USTDA, DoC, USDA, DoE, DoI, EXIM, OPIC/DFC) (*)	2018–2021 (Clean Asia EDGE since 2022)	Promote energy diversification and trade (focus on regulatory reforms, and private investments). Improve reciprocal energy trading relationships for exporting all forms of energy and related US technologies, including renewables (with Clean Asia EDGE, major focus on renewables and decarbonizations' objectives).	Indo-Pacific region
US–Japan Mekong Power Partnership	DoS–ENR	2019–ongoing	Linked to the Asia EDGE initiative. Focus on energy security, electricity networks, market integration, and regulatory reforms	Cambodia, Lao PDR, Thailand, Vietnam
Southeast Asia's Smart Power Program	USAID	2021–ongoing	Promote regulatory reforms, private investment, renewable energy, and regional market integration	ASEAN countries

Notes: (*) DoC = Department of Commerce; USDA = US Department of Agriculture; USTDA = US Trade and Development Agency; DoI = Department of Interior
Source: Author's elaboration from USAID websites available at https://www.usaid.gov/energy/programs (accessed 10–12 October 2023), and Department of State website, available at: https://www.state.gov/u-s-department-of-state-archive-websites/ (accessed 9–11 October 2023)

Table 5.5 Main USAID bilateral programmes for the power sector and renewable energy (since 2009)

Country	Programme	Years	Objectives
India	Partnership to Accelerate Clean Energy (*)	2009–ongoing	Capacity building, and regulatory reforms to promote the diffusion of renewable energy
Pakistan	Dam Rehabilitation Projects	2010–ongoing	Originally focused on rehabilitation of dams, later expanded to regulatory reforms in the power sector and promotion of renewable energy (including through private investments)
Kosovo	Energy Reform Project	2014–ongoing	Capacity building, and regulatory reforms for the power sector
Philippines	Building Low Emission Alternatives to Develop Economic Resilience and Sustainability/Energy Policy and Development Program	2014–ongoing	Capacity building, and regulatory reforms to promote the diffusion of renewable energy
Indonesia	Indonesia Clean Energy Development	2015–2020	Capacity building, and regulatory reforms to promote the diffusion of renewable energy
DRC	Power Sector Reform	2016–2019	Capacity building, and regulatory reforms for the power sector (including to favour the diffusion of renewable energy)
Mozambique	Supporting the Policy Enabling Environment for Development	2016–2019	Capacity building, and regulatory reforms for the power sector (including to favour the diffusion of renewable energy)
Kazakhstan	Energy Auctions for Kazakhstan's Green Economy	2017–ongoing	Capacity building, and regulatory reforms for promoting auctions for renewable energy
Bangladesh	Developing Renewable Energy in Bangladesh	2019–2021	Capacity building, and regulatory reforms for promoting renewable energy
Colombia	Reverse Energy Auction Design Support	2018–ongoing	Capacity building, and regulatory reforms for promoting auctions for renewable energy

Notes: (*) = programme developed in partnership with the DoC, DoS, DoE, EXIM, and OPIC/DFC
Source: Author's elaboration from USAID website, available at https://www.usaid.gov/energy/programs (accessed 10–12 October 2023)

A similar approach was followed by the MCC, which since 2010 had been increasing its engagement in the developing countries energy sector (Table 5.6). The MCC committed a larger amount of financial resources to support the development of electricity networks and power projects, as well as regulatory reforms. From 2010 to 2016, it committed more than $1.3 billion to these objectives, although not all its engagements focused on renewable energy (Table 5.6).

Through the DoS and the DoE, then, the Obama administration promoted bilateral dialogues and partnerships with (select) industrialized countries, the EU, and emerging economies (Table 5.7). In the latter case, the most important were with China (2008) and India (2009), as these led to upgraded versions of previous agreements that included a better focus on climate change and renewables. The 2008 framework agreement with China produced several other specific deals, including ambitious joint projects in the area of R&D, such as the US–China Clean Energy Research Center, which represented an innovative model of cooperation for the protection of intellectual property rights (Lewis 2023a).

Finally, along with these engagements—responding to a green soft power logic—the Obama administration mandated the OPIC and the EXIM to increase their support for the internationalization of US green business and the realization of renewable energy projects abroad. This shift towards economic tools, commercial diplomacy, and green economic power was consistent with the more interventionist approach that the Obama administration was implementing at home. It was developed, however, in parallel with its efforts to relaunch an international liberal agenda—a liberal agenda that, as we saw, failed.

Beginning in 2010, the OPIC increased its engagement with wind and solar PV projects, providing direct loans and guarantees to facilitate investment by US private companies in third countries (Figure 5.4). During Obama's second term, the OPIC further augmented this type of engagement (Figure 5.4). The EXIM had already established a programme for financing 'environmentally beneficial exports', including renewable energy equipment, in 1994. Like the OPIC, in this period the EXIM increased its funding for renewable energy projects, backing the export of US green technology, although on a much smaller scale (Figure 5.4). However, the EXIM's green commitments remained only a small fraction of its total portfolio. During the last years of the George W. Bush administration (2006–2007), the EXIM authorization for renewable energy exports as a percentage of total authorization averaged only 0.11% (Agarwal and Dupuy 2018). During the Obama years (2008–2016), this number significantly increased, but

Table 5.6 MCC energy engagements in third countries (since 2010)

Country	Year	Million $	Focus (specific energy programme)
Mongolia	2010–2015	40	Renewable energy and energy networks ('Energy and Environment Project')
Tanzania	2010–2015	200	Power sector regulatory reform, energy networks ('Energy Sector Project')
Indonesia	2013–2018	228	Renewable energy ('Green Prosperity Project')
Malawi	2013–2018	350	Power sector regulatory reform, energy networks
Liberia	2016–2021	257	Power sector regulatory reform, energy networks
Ghana	2016–2022	316	Power sector regulatory reform, energy networks ('Power Compact')
Benin	2017–2021	391	Power sector regulatory reform, energy networks ('Power Compact')
Nepal	2017–ongoing	398	Energy networks (including a link with India) ('Electricity Transmission Project')
Senegal	2018–ongoing	550	Power sector regulatory reform, energy networks ('Power Compact')
Kosovo	2018–ongoing	202	Regulatory reform, renewable energy
Burkina Faso	2020–2022	210	Power sector regulatory reform, energy networks

Source: Author's elaboration from the MCC website, available at: https://www.mcc.gov/where-we-work (accessed 2–3 November 2023)

it ultimately reached only 1.05% (ibid.). Even during this period, the EXIM's activities remained centred on more traditional business, such as air transportation and oil and gas projects. In 2009 these sectors accounted for 48% and 12% of the total EXIM financial exposure, respectively, and in 2016 these data were still 48% and 16%, respectively (about $42 billion and $14 billion; EXIM 2009, 2016). On the other hand, OPIC financing was more effectively directed towards renewable energy. During the period of maximum disbursement for such projects (2013–2015), about 30% of total OPIC financial commitments targeted renewables (OPIC 2013, 2015). It is worth noting that all OPIC and EXIM projects are accountable to high environmental and social standards and mandated to respect human rights, including workers' rights. These institutions operate within the OECD guidelines for export credit and aid, which the US itself contributed to developing. Hence, through its financial engagements, the US aims to spread such norms and principles globally.

Table 5.7 Main bilateral energy dialogues and partnerships established under Obama and Trump

Years	Initiative
2008	US–China Ten-Year Framework for Cooperation on Energy and Environment
2008	US–Indonesia Energy Policy Dialogue
2009	US–India Partnership to Advance Clean Energy
2009	US–EU Energy Council
2012	US–Brazil Strategic Energy Dialogue
2012	US–Korea Energy Policy Dialogue
2014	US–Ukraine Energy Partnership
2015	US–Brazil Climate Change Working Group
2016	US–Mexico Energy Business Council
2017	US–Japan Strategic Energy Partnership
2018	US–India Strategic Energy Partnership
2018	US–Poland Strategic Dialogue on Energy
2019	US–Brazil Energy Forum

Sources: Author's elaboration from the DoE and DoS websites (accessed 4–5 November 2023)

In total, during the Obama era, the OPIC and EXIM committed more than $5 billion to the realization of wind and solar projects abroad, which resulted in around 3,400 MW of newly installed capacity (Figure 5.4; Figure 5.5). However, EXIM operations were halted by Congress in 2015, which opposed the nomination of key members of its board of directors, de facto limiting its ability to authorize new financial transactions. Later on, during the first years of the Trump administration, the OPIC also significantly reduced its foreign engagements, including in the area of renewable energy (Figure 5.4). Activity would restart again in 2019, when the OPIC was transformed into the Development Financial Corporation (DFC). Yet, from 2009 to 2020, these institutions (especially the OPIC/DFC) promoted the realization of approximately 2,000 MW of wind and solar installed capacity in Asia, particularly in India, Pakistan, and Thailand (Table 5.8). In the case of India, these financial engagements were instrumental in reinforcing bilateral cooperation, implementing the 2009 Partnership to Advance Clean Energy, and helping the country meet its energy transition objectives. In several other countries, these financial engagements reinforced diplomatic and development efforts. Another 1,400 MW of installed capacity was promoted by the OPIC/DFC and the EXIM in Latin and Central America (particularly in Chile and Peru), and about 900 MW in Africa and the Middle East, mainly in Kenya, South Africa, Senegal, and Jordan (Table 5.8). Conversely, in the same period,

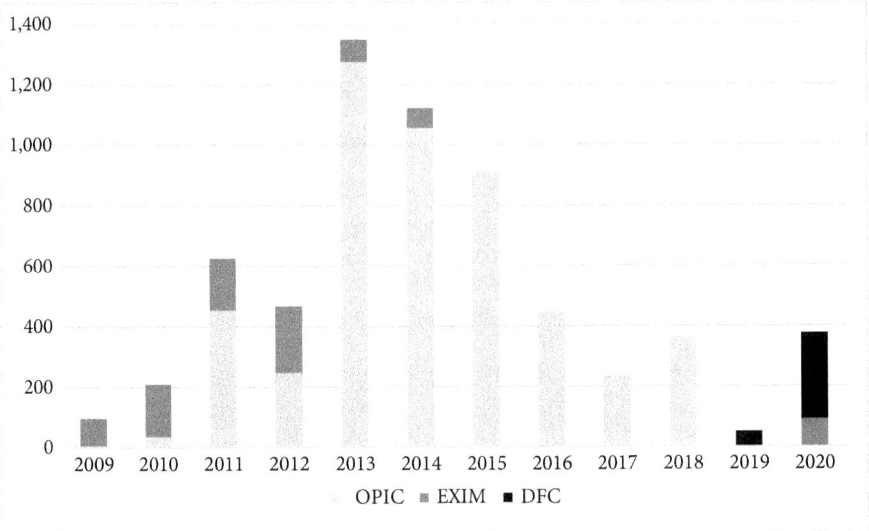

Figure 5.4 OPIC, EXIM, and DFC financial involvement in wind and solar PV projects abroad (in million $)

Source: For OPIC and EXIM's data from 2008 to 2018, author's elaboration from 'Bilateral Power Financing Dataset', available at: https://dataspace.princeton.edu/handle/88,435/dsp01pv63g332x (accessed 4 October 2023). For DFC, author's elaboration from DFC's projects database, available at: https://www.dfc.gov/what-we-do/active-projects (accessed 11 December 2023). For the 2020 EXIM data, see https://www.exim.gov/news/exim-board-approves-915-million-export-financing-for-renewable-energy-rural-electrification (accessed 12 December 2023)

foreign direct investment by US companies for such projects was minimal (Chen, Gallagher, and Mauzerall 2021). A partial exception was in Latin and Central America, where US foreign direct investment targeted Brazil (596 MW of wind installed capacity), Mexico (461 MW of wind installed capacity and 210 MW of solar PV), and Peru (110 MW of wind installed capacity; ibid.).

With regard to Africa, the main initiative put forward by the Obama administration was Power Africa. Launched in 2013, it aimed to double access to electricity in sub-Saharan Africa by increasing installed generation capacity (with a target of 30,000 MW by 2030) and promoting electricity networks. It represented the flagship project of the Obama era's engagements with the region, and it was also intended to counter China's BRI. Power Africa was designed as a government-led partnership; it was coordinated by the USAID and involved several US governmental agencies and financial institutions (Table 5.4). However, it also included a diverse and large coalition of more than 140 public and private sector partners, such as bilateral and multilateral development banks (e.g., AfDB, World Bank,

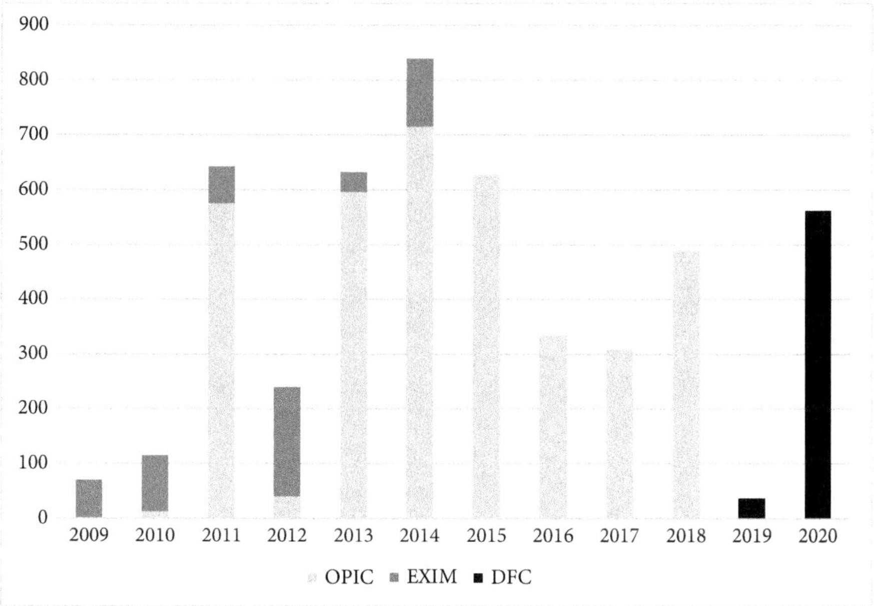

Figure 5.5 Solar PV and wind projects (in MW of installed capacity) promoted abroad thanks to OPIC, EXIM, and DFC financial support

Source: For OPIC and EXIM's data from 2008 to 2018, author's elaboration from 'Bilateral Power Financing Dataset', available at: https://dataspace.princeton.edu/handle/88,435/dsp01pv63g332x (accessed 4 October 2023). For DFC, author's elaboration from DFC's projects database, available at: https://www.dfc.gov/what-we-do/active-projects (accessed 11 December 2023). In 2020 EXIM committed $91 million for exporting US solar equipment to Senegal for the realization of small projects of rural electrification

the Development Bank of Southern Africa, the Eastern and Southern Africa Trade and Development Bank), foreign donors (e.g., the EU, Canada, Denmark, France, Israel, Japan, South Korea, Norway, Sweden, and the UK), civil society organizations, and more than one hundred private companies. The OPIC alone, from 2014–2017, committed around $2.4 billion to Power Africa, on top of which the EXIM committed further. These resources sustained the internationalization of US green business, and these institutions should be considered part of Washington's commercial diplomacy armamentarium. However, Power Africa mainly works through those mechanisms of mobilization, public–private partnerships, and network diplomacy that help to create a multi-actor coalition and that characterize catalytic power. By the end of the Obama era, Power Africa had mobilized more than $14.5 billion for energy projects in many African countries, thanks to the involvement of several public and private partners (Power Africa 2017). Moreover, it coupled investment efforts with capacity building and regulatory and market

Table 5.8 Solar PV and wind projects (in MW of installed capacity) financed by OPIC, EXIM, and DFC in various regions and countries (from 2009 to 2020)

	Wind (MW)	Solar PV (MW)	Total (MW)
Asia	Pakistan (299.1), India (252), Indonesia (75)	India (340.6), India (555) (DFC), Thailand (520)	2041.7
Latin and Central America	Peru (238.3), Mexico (148.5), Honduras (126), Uruguay (100), Jamaica (34), St. Christopher-Nevis (5.4)	Chile (452.8), El Salvador (157), Peru (99), Panama (32.5), Jamaica (20), Mexico (15.5)	1429
Africa and Middle East	Kenya (410), Senegal (158.7)	South Africa (160), Jordan (101.9), Zambia (55), Malawi (37) (DFC), Burundi (7.5) (DFC), Tanzania (5)	935.1
Total	1847	2558.8	4405.8

Notes: (DFC) = funding by DFC. OPIC also financed solar projects in Israel (110 MW) and a wind project in Ukraine (230 MW)
Source: For OPIC and EXIM's data from 2008 to 2018, author's elaboration from 'Bilateral Power Financing Dataset' available at: https://dataspace.princeton.edu/handle/88,435/dsp01pv63g332x (accessed 4 October 2023). For DFC author's elaboration from DFC's projects database available at: https://www.dfc.gov/what-we-do/active-projects (accessed 11 December 2023)

reforms. In February 2016, Congress, with a bipartisan agreement, passed the Electrify Africa Act, which institutionalized Power Africa through legislation and signalled that expanding electricity access in sub-Saharan Africa was a long-term US foreign policy priority. However, Power Africa did not focus only on solar and wind power. Even during the Obama era, natural gas and hydropower were important components of this initiative (Table 5.9). In addition, the majority of the solar and wind projects were concentrated in a small group of countries, particularly South Africa, which was the main beneficiary in terms of newly installed capacity (South Africa is also the largest US trade partner in Africa), Kenya, Senegal, and Zambia (Table 5.9).

As illustrated, with the arrival of Trump, US foreign energy policy shifted towards fossil fuels. Many of the cooperative efforts with China on climate and renewable energy that had been developed under Obama were halted, and no new bilateral agreements were signed by Washington and Beijing (Lewis 2023). Conversely, the Trump administration, particularly at the beginning of its mandate, focused on increasing fossil fuel exports to China. At the same time, some building blocks of the US external green action were not discontinued, although the new administration put a stronger

Table 5.9 Power Africa Initiative: installed capacity (in MW) by technology, and main countries targeted (in MW)

Technology	Years (*)			Total (MW)	Main targeted countries (MW)
	2013–2017	2017–2020	2020–2023		
Solar PV (MW)	628	1350	710	2688	South Africa (1200), Senegal (80), Zambia (88), Malawi (60), Mozambique (40), Kenya (40)
Wind (MW)	810	1636	60	2506	South Africa (2000), Kenya (310), Senegal (158)
Hydro (MW)	1030	542	11	1583	Nigeria (500), Zambia (120), Liberia (90), Rwanda (80)
Natural Gas (MW)	4315	45	2247	6607	Nigeria (2500), Tanzania (580), Ghana (550)
Total (MW)	6783	3573	3028	13,384	

Notes: (*) = Financially closed projects for each time period. The table does not cover (small) renewable energy projects for off-grid solutions
Source: Author's elaboration from Power Africa Annual Report, various years, available at: https://www.usaid.gov/powerafrica/annualreport (accessed 2–3 December 2023)

emphasis on market reforms and the promotion of private investments than on decarbonization objectives.

The Obama administration had embraced the post-2015 UN development agenda. Trump did not formally reverse this choice, but his administration downplayed several UN multilateral initiatives, including this one. It reneged on the US financial commitment to the Green Climate Fund. Moreover, it sought to shrink US foreign aid and development policy, including by proposing drastic budget cuts to the USAID. These proposals, however, encountered effective resistance within Congress and the US foreign aid bureaucracy and from NGOs and actors from the development community (e.g., Regilme 2023). The main USAID programmes initiated in previous years endured, and new ones were launched at both global and bilateral levels (Table 5.4; Table 5.5). These include Scaling Up Renewable Energy, Energy Auctions for Kazakhstan's Green Economy, and Reverse Energy Auction Design Support for Colombia, which focused on capacity building and the diffusion of tools for renewables able to attract private investments such as auctions. Through the auctions organized with the USAID's assistance, nearly

15,000 MW of new solar and wind capacity was installed between 2016 and 2020 in multiple countries in the Global South (USAID 2023).

The Trump administration continued the Power Africa initiative. In 2018, however, it launched Power Africa 2.0, refocusing the original project towards distribution and transmission networks and downsizing the official references to renewable energy and climate goals. Yet, during the Trump years, more financial agreements for solar and wind projects were closed than in the previous period (Table 5.9). With regard to Africa, the Trump administration, in 2019, launched an additional multi-agency initiative—i.e., Prosper Africa—that involved leading US government agencies and financial institutions, such as the USAID, DoS, DoE, DFC, EXIM, and MCC. Prosper Africa was intended to increase trade and investment between the US and African countries by supporting private business, as well as counter China's rise in the continent. Although not centred on renewables, some of the projects realized by US private companies and investors targeted (small) solar power installations and electricity grid expansion.

A similar approach was also followed in Latin and Central America. In 2019, the Trump administration launched Americas Grow (or America Crece) as a multi-agency initiative—involving the DoS, DoE, USDT, DoC, USAID, United States Trade and Development Agency (USTDA), and OPIC—for facilitating private sector investments for infrastructure in the region and countering the growing Chinese role in this field. The US signed a Memorandum of Understanding with Argentina, Brazil, Chile, Colombia, Ecuador, El Salvador, Jamaica, and Panama. However, this initiative was neither backed by additional financial resources nor focused on renewable energy; it was a repackage of existing programmes (e.g., the 2014 Caribbean Energy Security Initiative) and engagements; for example, under this framework the OPIC continued to provide support for solar projects in Chile.

In these years, the MCC also continued its engagement with the Global South, promoting regulatory reforms in the power sector and investments in energy networks (Table 5.6). New bilateral strategic partnerships and dialogues were established between the US and Japan (2017), India (2018), and Poland (2018; see Table 5.7). Cooperation with Poland focused on nuclear energy, energy security, and expanding export of US LNG. Fossil fuels, LNG, and energy security were also key components of the new partnerships with Japan (along with nuclear energy) and India. However, in these two cases, cooperation extended to the area of renewable energy. Conversely, unlike the US–Brazil Climate Change Working Group launched under Obama, the 2019 US–Brazil Energy Forum adopted a strong focus on fossil fuels.

Strategic energy partnerships with India and Japan were part of the broader strategy of the Trump administration to address the rise (military, economic, and diplomatic) of China in the Indo-Pacific (US Department of State 2019). This included the relaunch, in 2017, of the Quadrilateral Security Dialogue (Quad)—comprising Australia, India, and Japan—improving bilateral relations with several other countries in the region (e.g., South Korea, the Philippines, Thailand), and US engagement with the ASEAN. This strategy, which aimed to promote market-based economic systems, private sector finance, and an open investment environment, also had an energy component represented by the Asia EDGE (Table 5.4). This multi-agency initiative, launched in 2018, did not have an explicit focus on energy transition or climate mitigation, instead prioritizing energy security, market reforms, and energy trade. Yet, it encompassed an 'all-of-the-above approach' to energy security: from fossil fuels and nuclear energy to renewables and sustaining the export of US green technologies.

Moreover, in 2019 the US, Australia, and Japan—and their respective development financial institutions—launched the Blue Dot Network initiative as a framework for promoting global investments in infrastructure that could function as an alternative to China's BRI and its state-led and non-OECD model of intervention. This framework was designed as a global standard-setter certification scheme—aligned with the OECD and G7 guidelines—to improve the quality and sustainability of projects and better attract private capital for their realization.

This latter effort followed a regulatory power approach logic, although it was not a proper manifestation of a green regulatory power strategy by Washington. This was also the case for the USMCA, which replaced the NAFTA in 2020. This new trade agreement negotiated by the Trump administration provided for an 'environment chapter' with commitments to implement key multilateral environmental agreements to which the US was a party. Hence, it did contain provisions to address climate change or mitigation policies (the USMCA did not include references to the UNFCCC or the Paris Agreement). However, like the NAFTA, it maintained provisions for increasing regulatory cooperation and energy market integration (which, as noted, has been especially developed between the US and Canada). Moreover, the USMCA contains provisions to facilitate cross-border trade and investment in renewable and low-carbon technologies as well as hydrogen and hydrogen-related technologies (the US originally protected solar equipment, but in 2022 Canada successfully leveraged the USMCA to eliminate tariffs on imports from Canada).

Conversely, the Trump administration initially sought to halt the work on renewables conducted by the OPIC and EXIM. As explained, EXIM operations had already been de facto limited in 2015; the Trump administration's first budget request, in June 2017, proposed eliminating the OPIC altogether. This position was soon revised, as the administration was preparing its new National Security Strategy, which emphasized rising competition with China. In this context, it began to frame the OPIC and the US development policy apparatus as a means to provide countries in the Global South with an alternative to China's state-led, so-called 'debt diplomacy' financing (Morgenstern et al. 2021). Hence, in 2019 the Trump administration endorsed an emerging bipartisan proposal—the Better Utilization of Investments Leading to Development Act—for establishing a more muscular and coordinated approach that upgraded the OPIC into a new development institution, i.e., the DFC (ibid.). The DFC was created by merging the OPIC with the USAID's Development Credit Authority. Its total investment limit was increased up to $60 billion, doubling the previous OPIC threshold. In addition, the DFC was provided with the ability to make equity investments along with debt financing, and it was mandated to work in closer cooperation with the DoS and USAID to advance US foreign policy objectives. Despite the global crisis spurred by the Covid-19 pandemic, in 2019–2020 the DFC made new investments to promote renewable energy projects abroad (Figure 5.4; Figure 5.5). In 2020 alone, the DFC contributed to solar projects in India for 555 MW of installed capacity (Table 5.8).

In this period, competition with China also led to relaunching the EXIM with an explicit green focus. In 2019 Congress reauthorized funding for the EXIM and mandated that the bank establish the Program on China and Transformational Exports. This new programme, which was responsible for 20% of the EXIM's total financing authority (around $27 billion out of a total of $135 billion authorized by Congress until 2026), aimed to provide loans, guarantees, and other financial tools that were competitive with China's foreign assistance and could support the internationalization of US companies and the diffusion of US technological standards in ten specific transformative industries, among which were renewable energy and energy storage.

Finally, the Trump administration strengthened the external dimension of the US approach to critical minerals, both through diplomatic engagements (bilateral, minilateral, and multilateral) and through the DFC. In 2019 it established the Critical Minerals Mapping Initiative with Australia and Canada in order to share data, coordinate policy, and identify new sources of supply. The same year, the US–Canada Joint Action Plan on Critical Minerals Collaboration was issued to foster cooperation with the country that was already the main supplier of many of the minerals deemed critical by Washington (thirteen of the thirty-five minerals included in the 2019 list). In 2019 the DoS also established—in partnership with Australia, Botswana, Canada, and Peru—the Energy Resource Governance Initiative (ERGI), a multilateral effort to advance mining sector governance, promote resilient and secure mineral supply chains, and encourage a level playing field. This initiative, which had an explicit focus on those minerals critical for renewable energy and low-carbon technology, was then joined by the Philippines, Argentina, Brazil, the DRC, Namibia, and Zambia. Finally, in 2019 the DFC made its first equity investment (worth $25 million) in TechMet—a private investment platform that comprises several US-based companies—to support the development of a mining facility in Brazil (in the State of Piauí), which will produce nickel and cobalt suitable for lithium-ion batteries.

Greening a (declining) hegemonic power

The renewable energy complex exhibited a certain resilience under Trump, who was not able to dismantle many internal and external components of the US green approach. This was also the result of the growing rivalry with China. However, it was only with Biden that Washington developed a comprehensive green foreign energy policy and a more defined green power strategy that was embedded in its wider global strategy. As mentioned, the first act of the Biden administration was to recommit the US to the Paris Agreement. In 2021 the USDT released its new Fossil Fuel Energy Guidance for Multilateral Development Banks, which advocated investments abroad that prioritize renewable energy and decarbonization objectives in line with the Paris Agreement while opposing support for coal and oil projects as well as for upstream natural gas projects (but not for midstream and downstream natural gas projects). These were necessary steps to relaunch Washington's multilateral climate diplomacy, which ran in parallel with the relaunch of minilateral and bilateral channels of climate cooperation; in the latter case, especially with major emitters such as China and India (cooperation with India extended to several other areas). This manifestation of US green soft power mirrored the approach followed under Obama, with minilateral and bilateral engagements intended to reinforce the UN climate regime. In April 2021, the Biden administration organized the Leaders' Summit on Climate, gathering industrialized countries and emerging economies, such as China, India, Brazil, and South Africa. On this occasion, Biden submitted the US NDC to the UNFCCC, several countries made new pledges, and others confirmed or reaffirmed their climate targets. The summit also resulted in the establishment of the US–India Climate and Clean Energy Agenda 2030 Partnership. This partnership built on preexisting agreements (e.g., the 2018 US–India Strategic Energy Partnership) but enhanced their green component. In April 2021, the US also launched the Net-Zero Producers Forum (with Canada, Norway, Qatar, and Saudi Arabia) as a platform for policy dialogue among oil- and gas-producing countries to support the implementation of the Paris Agreement and net-zero emissions targets. Later, in September 2021, Biden convened the Major Economies Forum on Energy and Climate, which was revitalized under his administration. All these engagements were instrumental in preparing for the COP 26 in Glasgow.

At the COP 26, apart from renewed dialogue with China, the US facilitated the establishment of the Global Methane Pledge. It also signed the Glasgow Statement—along with the EIB and several European countries—which aimed to end new international public support for the 'unabated' fossil fuel energy sector by the end of 2022 and prioritize investments in the energy transition. At Glasgow, the DoE and NREL launched another minilateral initiative, i.e., the Net-Zero World Initiative, for accelerating the diffusion of renewables and the decarbonization of energy systems (Argentina, Chile, Egypt, Indonesia, Nigeria, Singapore, Thailand, and Ukraine joined this project). The DoS-ENR also launched a new initiative (i.e., the Clean Energy Demand Initiative) to increase private companies' participation in renewable energy projects around the world. This reflected a wider shift in DoS foreign energy engagements under Biden. Indeed, although attention to traditional concerns related to energy security (for the US and its allies) and market reforms did not disappear, the DoS began to more explicitly link these long-standing issues with a focus on decarbonization and the energy transition.

At the COP 27 in Sharm el-Sheikh in 2022, the US endorsed the idea of a losses and damage fund. In the run-up to the COP 28, the US organized, in April 2023, another summit of the Major Economies Forum on Energy and Climate. Later, in November 2023, the US and China held a summit at Sunnylands in California, where Biden and Xi Jinping jointly reasserted their support for the UN multilateral climate agenda. Moreover, they committed to 'accelerate renewable energy deployment in their respective economies' and to contribute to the G20's goals to 'triple renewable energy capacity globally by 2030' (US Department of State 2023a). Again, these US diplomatic engagements were important to the results of the COP 28 in Dubai, where

Washington sustained final agreements envisaging a 'transitioning away' from fossil fuels. At the COP 28, the US, along with the EU (but not China), also signed the Global Renewables and Energy Efficiency Pledge, committing to triple the world's installed renewable energy generation capacity by 2030.

This (bilateral, minilateral, and multilateral) diplomacy was complemented by increased US climate financing. In 2021 the Biden administration launched the US International Climate Finance Plan, committing to work with Congress to scale up US international public climate financing to over $11 billion annually by 2024 (four times more than the highest previous levels provided by the US). From 2021 to 2022, total US climate financing increased from $1.5 billion to $5.8 billion, of which $2.8 billion was for renewable energy (US Department of State 2023b). Part of this funding was devoted to relaunching the US's role in multilateral efforts. For example, in 2022 the US pledged $600 million to the Global Environment Facility, and in 2023 it allocated $1 billion for the Green Climate Fund.

In addition, the Biden administration elevated climate change and energy transition objectives in US bilateral development policy and finance. In 2021, the USAID highlighted climate action as a major agency priority, including the goals of helping partner countries to implement their NDCs under the Paris Agreement and accelerating the deployment of renewable energy. As we saw, USAID actions in this latter field, which were mostly based on capacity building, technical assistance, and limited financial support, continued even during the Trump presidency. However, now they were firmly prioritized in the USAID engagements, which eventually incorporated other energy transition priorities as well, such as decarbonization in the industrial sector and improving the management of critical minerals supply chains (USAID 2022). In the area of renewables, the USAID continued to rely on existing programmes as well as launched new ones, such as the 2021 Southeast Asia's Smart Power Program for promoting regulatory reforms in the power sector, regional market integration, and the deployment of renewables (Table 5.4). Moreover, it established new vehicles for mobilizing private funds for climate action and energy transition, such as the Climate Finance for Development Accelerator (2022) and the Blended Finance for the Energy Transition initiative (2023) in cooperation with the DoS.

The DFC, as mandated by the Biden administration in 2021, also updated its development strategy to prioritize climate action and renewable energy, and the organization's climate financing grew from less than $500 million in 2020 to over $3.7 billion in 2023 (US Department of State 2023b). However, DFC engagements were also directed towards more strategic goals, such as improving the diversification of low-carbon supply chains, especially in the industrial and critical minerals sector (see below).

The MCC's work on power sector reforms continued under Biden; in addition, the agency increased attention to climate action. In 2022 it launched, in cooperation with the USAID, the Climate Finance+ programme to promote green bonds and other climate-related financing mechanisms in developing countries (initially targeting Indonesia, Mozambique, and Zambia). The MCC, the USAID, and the DFC continued to be involved in the Power Africa initiative as well, which was refocused on climate action and energy transition. From 2021 to 2023, the Biden administration invested almost $300 million to support Power Africa. However, even due to the crisis caused by Covid-19, the impact in terms of new renewable installed capacity was lower than in previous years (Table 5.9). Power Africa was flanked by new green programmes, such as the US–Africa Clean Tech Energy Network, which worked in partnership with Prosper Africa (also maintained by Biden) to help connect US and African green energy companies and increase private investment in the continent. This renewed focus on climate action and a 'just energy transition' in Washington's engagement in Africa was confirmed in the 2022 US Strategy Toward Sub-Sahara Africa, which also aimed to counter China's influence by championing an approach based on 'high-standards, transparent investments, and (shared) democratic values' (The White House 2022c).

This mixture of energy transition and more strategic objectives—and of diplomatic and economic tools—is also visible in US support for the creation, along with the EU and other Western allies, of a set of Just Energy Transition Partnerships between 2021 and 2023 (see Chapter 4). Indeed, these Western-led (minilateral) initiatives provide a more focused approach than the UNFCCC multilateral model (Jakob and Martini 2023). They combine diplomacy and conditional financing to promote the deployment of renewables, as well as the diffusion of some norms and values for shaping the energy transition. At the same time, they are part of the G7's geopolitical strategy to offer an alternative to Chinese development financing in the Global South.

The Biden administration began to promote this strategy in 2021, when the G7 endorsed the Build Back Better World Partnership, a climate-friendly global infrastructure initiative aligned with Western-led financing standards—an alternative to the Chinese BRI. During the following 2022 G7 Summit, amid the beginning of the war in Ukraine and the further deterioration of West–China relations, this initiative was upgraded into the Partnership for Global Infrastructure and Investment (PGII), which itself built on the principles of the Blue Dot Network. The Biden administration aimed to mobilize $200 billion for the PGII by 2027 through public funding and leveraging private sector investments. Together with G7 partners' programmes, such as the EU's Global Gateway and the UK's British International Investment, the goal was to mobilize $600 billion by 2027. During the 2021 and 2022 G7 Summits, Washington also called for cooperation among Western countries and 'like-minded' partners to reduce dependencies in critical low-carbon supply chains. This issue was confirmed at the 2023 G7 Summit held in Japan, which reasserted the emphasis on promoting sustainability and high environmental, social, and governance standards for these supply chains.

The Biden administration specifically mandated that the DFC and EXIM sustain the PGII. To achieve this goal, in 2021 the EXIM created a new division, i.e., the Office of Global Finance Development. Like the DFC, the EXIM was mandated to implement Biden's green agenda abroad, orienting its priorities towards renewable energy, sub-Saharan Africa, and the sectors covered by the Program on China and Transformational Exports (which also comprises renewable energy). After a very low output in 2021, in 2022–2023 the EXIM and DFC gradually increased their financial support for renewable energy projects abroad (Table 5.10). In 2023 the EXIM provided a $900 million loan to the Angolan government for a large solar project linked to the PGII and the Program on China and Transformational Exports. Moreover, in 2022–2023 the DFC provided financing and equity investments for the diversification of low-carbon supply chains, targeting projects for critical minerals (in Brazil, South Africa, and Mozambique) and solar energy manufacturing expansion in India (Table 5.10). In 2023 the Biden administration also facilitated a strategic partnership between TechMet and Life Zone Metals to work with the Tanzanian government to open a facility to process nickel and other critical minerals extracted in the country, targeting delivery of battery-grade nickel to the US and global market as soon as 2026 (The White House 2023c).

Along with the deployment of US economic tools and commercial diplomacy, the Biden administration continued the diplomatic initiatives in the area of critical minerals that Trump had initiated, such as the ERGI, and further improved cooperation with Australia and Canada. The US enhanced cooperation with Canada regarding the implementation of the 2020 US–Canada Joint Action Plan on Critical Minerals, as well as in other areas of low-carbon technology supply chains, such as the harmonization of standards for EV charging stations. In 2022, through the Defense Production Act, $250 million was allocated to US and Canadian companies jointly working in the mining and processing of critical minerals for EVs and stationary batteries. In addition, in 2022 the US launched a new minilateral

Table 5.10 EXIM and DFC funding for renewable energy, critical minerals and manufacturing projects abroad (2021–2023)

Year	EXIM	DFC
2021	• $8.7 million for a 10 MW solar project in Guatemala	• —
2022	• $52 million for a 53 MW solar project in Honduras	• $50 million for a 300 MW solar project in Egypt • $25 million for a 20 MW solar project in Malawi • $20 million for 25 MW solar project in India • $500 million for the construction of a 3.3 GW thin film solar panels manufacturing facility in India • $30 million (equity) for TechMet's operations in the Brazilian nickel and cobalt mining sector
2023	• $900 million for a 500 MW solar project in Angola	• $192 million for 400 MW solar projects in Ecuador • $425 million for 4 GW solar PV cells, and 4 GW solar PV modules manufacturing facilities in India • $50 million (equity) for TechMet's REEs projects in South Africa • $150 million for the expansion of a graphite mining and processing facility in Mozambique

Notes: From 2021 to 2023, DFC also provided finance and equity investments worth $237 million for different regional funds supporting renewable energy in Africa and Asia
Source: Author's elaboration from EXIM Annual reports (available at: https://www.exim.gov/news/reports), DFC Projects Database (available at: https://www.dfc.gov/what-we-do/active-projects), and press sources

initiative, i.e., the Mineral Security Partnership, which included Australia, Canada, Finland, France, Germany, India, Italy, Japan, Norway, South Korea, Sweden, the UK, and the EU. This partnership focused explicitly on cooperation on those minerals and metals most relevant for the energy transition, such as lithium, cobalt, nickel, REEs, copper, and graphite. Finally, in 2023 Washington established a US–Mongolia–South Korea trilateral dialogue to advance information exchange and collaboration on critical minerals supply chains (Mongolia is rich in minerals, especially copper and REEs).

The incorporation of green foreign energy policy objectives into the US global strategy is further visible in Washington's new regional and bilateral engagements. In 2022 the Biden administration issued the new US Indo-Pacific Strategy; as during the Trump years, it aimed to address China's rise in the region. However, the reissued version emphasized diplomatic and financial cooperation to tackle climate change, accelerate decarbonization and energy transition, and improve the security of low-carbon technology supply chains (The White House 2022d). These objectives were incorporated into the existing regional energy programmes. In 2022 the EDGE Asia initiative was transformed into the Clean EDGE Asia, and even the US–Japan Mekong Power Partnership was redirected towards renewable energy development, battery systems, and EV infrastructures. In addition, as part of its Indo-Pacific strategy, in May 2022, the Biden administration created a new initiative, i.e., the Indo-Pacific Economic Framework for Prosperity (IPEF).

The IPEF included the five regional 'treaty allies' of the US—i.e., Australia, South Korea, Japan, the Philippines, and Thailand, which have defence treaties in place with Washington—and its other Indo-Pacific partners (Table 5.11). Like the other regional and bilateral initiatives developed under Biden, the IPEF did not aim to reduce trade barriers, as did Obama's TPP or the Regional Comprehensive Economic Partnership Agreement, which included China (see Chapter 3). Its goal was to improve dialogue, cooperation, and coordination in a flexible way in four pillar areas, among which low-carbon technology and critical minerals supply chains, renewable energy, decarbonization, and infrastructure (e.g., electricity grids) occupied a central position (Table 5.11). To support investments, specific platforms have been established under the IPEF in connection with the PGII. Washington upgraded cooperation with ASEAN with the launch, in 2022, of the US–ASEAN Comprehensive Strategic Partnership (Table 5.11). This partnership also emphasized cooperation on energy transition, renewable energy and low carbon technology, and climate action (e.g., the US increased support for the ASEAN Centre for Climate Change). Additional focus was devoted to promoting US green technologies and standards, developing an EV regional ecosystem (with the US–ASEAN Electric Vehicle Initiative), and enhancing connectivity. The US–ASEAN Platform for Infrastructure and Connectivity was established in order to facilitate investment in regional infrastructure projects under the auspices of the PGII.

Other regional initiatives with a green foreign energy policy dimension and financial mechanisms for mobilizing private investments were subsequently created for Latin and Central America and the Middle East: the Americas Partnership for Economic Prosperity (APEP) and the India–Middle East–Europe Economic Corridor (IMEC; Table 5.11). These were also intended to counter the growing Chinese influence in these regions. The latter, in particular, saw the involvement of other Western allies and the

Table 5.11 Main regional initiatives launched by the Biden administration, their green foreign energy policy dimension and financial framework

Initiative	Year	Countries	Green foreign energy policy dimension	Financial framework
Indo-Pacific Economic Framework for Prosperity (IPEF)	2022 (*)	Australia, Brunei, Fiji India, Indonesia, Japan, South Korea, Malaysia, New Zealand, Philippines, Singapore, Thailand, and Vietnam	• Accelerate the development and deployment of renewable energy technologies • Enhance electricity connectivity, electric grids, and mini-grids • Supply chains diversification and resilience for critical minerals, and low-carbon technology • Scaling up zero emission vehicles and their associated infrastructure • Decarbonization of industry and climate actions • Cooperation on green technologies and standards	• PGII–IPEF Investment Accelerator, IPEF Clean Economy Investor Forum, IPEF Catalytic Capital Fund • IPEF Project Preparation Facility (managed by the US Trade and Development Agency)
US–ASEAN Comprehen-sive Strategic Partnership	2022	ASEAN members	• Develop an integrated EVs ecosystem in South East Asia (e.g., US–ASEAN Electric Vehicle Initiative) • Promote climate action and the energy transition • Enhance the region's connectivity, and invest in sustainable infrastructures • Cooperation on green technologies and standards	• PGII
Americas Partnership for Economic Prosperity (APEP)	2022	Barbados, Canada, Chile, Colombia, Costa Rica, Dominican Republic, Ecuador, Mexico, Panama, Peru, Uruguay	• Accelerate the development of renewable energy • Enhance decarbonization and climate action • Cooperation on green technologies and standards • Low-carbon supply chains diversification and resilience	• Joint investment platform between DFC and the Inter-American Development Bank (IDB)
India–Middle East–Europe Economic Corridor (IMEC)	2023	India, Saudi Arabia, UAE, France, Germany, Italy, EU	• Facilitate the development and export of renewable energy and hydrogen • Develop transnational electricity grids • Low-carbon supply chains diversification and resilience	• PGII

Notes: All these initiatives also emphasize promoting good quality jobs, environmental criteria and transparency for infrastructure, and private investments; (*) = several cooperation agreements in the context of the IPEF were finalized in 2023, including the IPEF Clean Economy Agreement, the IPEF Supply Chain Agreement, and IPEF Critical Minerals Dialogue, the IPEF Investment Accelerator, the IPEF Clean Economy Investor Forum, the IPEF Catalytic Capital Fund, the IPEF Project Preparation Facility
Sources: Author's elaboration from US government's website

EU. It was launched in September 2023 during the 2023 G20 Summit in New Delhi (India was a crucial partner in the project).

These regional initiatives provide for new institutional venues and diplomatic channels for cooperation as well as economic tools and mechanisms for mobilization and for creating multi-actor coalitions in line with catalytic power perspective. Moreover, they were complemented by bilateral engagements (Table 5.12). This was particularly the case in the Indo-Pacific and the Middle East. As already noted, in 2021 the Biden administration had improved cooperation with India on climate action and the energy transition. In 2022 the US–Japan Clean Energy and Energy Security Initiative and the US–Japan Climate Partnership were established to foster cooperation in the areas of climate change, renewable energy, and low-carbon technology, such as energy storage and batteries, smart grids, hydrogen, and critical minerals supply chains. A Net-Zero Technology Acceleration Partnership was launched with Australia in 2022. Then, in 2023, the Australia–US Climate, Critical Minerals and Clean Energy Transformation Compact was established to elevate 'climate and clean energy as a central pillar of the Australia–United States alliance' (The White House 2023d). This framework did not aim to increase trade liberalization (the US and Australia have had a Free Trade Agreement since 2005). Rather, its goals were to coordinate

Table 5.12 Main bilateral energy transition-related dialogues and partnerships established under Biden

Year	Initiative
2021	US–India Climate and Clean Energy Agenda 2030 Partnership
2021	US–Egypt Climate Working Group and High Level Joint Economic Commission
2022	US–Brazil Clean Energy Industry Dialogue
2022	US–Japan Clean Energy and Energy Security Initiative, US–Japan Climate Partnership
2022	US–Australia Net-Zero Technology Acceleration Partnership
2022	US–UAE Partnership to Accelerate Transition to Clean Energy
2022	US–Philippines Energy Policy Dialogue
2023	US–Australia Critical Minerals and Clean Energy Transformation Compact
2023	US–Vietnam Comprehensive Strategic Partnership
2023	US–Indonesia Comprehensive Strategic Partnership
2023	US–Thailand Communiqué on Strategic Alliance and Partnership
2023	US–Saudi Arabia MoU for Establishing Intercontinental Green Transit Corridors

Sources: Author's elaboration from US government's website

industrial policy, upgrade cooperation in the areas of critical minerals supply chains, renewable energy, and low-carbon technology (with a focus on batteries), as well as with related international standards, and to develop clean hydrogen in the respective countries and across the Indo-Pacific.

In this region, the Biden administration also targeted Vietnam and Indonesia: comprehensive strategic partnerships were signed with these countries in 2023. The partnership with Indonesia prioritized supporting the diffusion of renewable energy and decarbonization and promoting both grid interconnection projects—e.g., between Indonesia and Malaysia—and technical and regulatory cooperation on critical minerals. With Vietnam, the focus was on decarbonization, renewable energy, batteries, and critical minerals supply chains (Vietnam has potential in the area of REEs). Bilateral (green) cooperation also increased with Thailand, which had joined the Net Zero World Initiative. In 2022, the US and Thailand issued a Communiqué on Strategic Alliance and Partnership, which included closer cooperation in the areas of climate change, renewable energy, EVs, batteries, and critical supply chains.

In the Middle East, US actions targeted major allies and partners (e.g., Egypt) and the Gulf's oil-rich monarchies, Saudi Arabia and the UAE. Whereas, during the Trump administration, energy cooperation with Egypt focused on the exploitation of Eastern Mediterranean gas resources, under Biden it shifted towards climate change and renewables. In 2021, during the first meeting of the US–Egypt Strategic Dialogue held by the new administration, a joint US–Egypt Climate Working Group was established and a high-level Joint Economic Commission was launched to increase cooperation in the areas of renewables, sustainable infrastructure, and low-carbon supply chains. With the UAE, in 2022 Washington established the Partnership to Accelerate Transition to Clean Energy, the goals of which were to 'mobilize $100 billion in financing, investment, and other support and to deploy globally 100 gigawatts of clean energy by 2035' (The White House 2022e). This partnership also provided for cooperation on green innovation, industrial decarbonization, hydrogen, and critical minerals supply chains. With Saudi Arabia, the US signed a Memorandum of Understanding for Establishing Intercontinental Green Transit Corridors (Table 5.12). This agreement, linked to the IMEC, envisages the promotion of renewable energy, transnational electricity links, and clean hydrogen corridors between Asia and Europe. Washington also started a dialogue with Saudi Arabia on investments in the area of critical minerals supply chains in Africa. In this vein, in 2022 the US signed a Memorandum of Understanding with the DRC and Zambia (the largest and second largest cobalt producers in Africa, respectively) to sustain their joint efforts to develop critical minerals supply chains for EVs and batteries.

Finally, in Latin and Central America, the Biden administration leveraged the US–Brazil Energy Forum, established under Trump, to strengthen cooperation in the areas of renewable energy, electricity grids, green hydrogen, and related regulatory frameworks and standards, and in 2022 the two countries established the US–Brazil Clean Energy Industry Dialogue. Argentina and Chile had already joined the Net-Zero World Initiative by 2021, but it was not until 2023 that the US and Chile launched bilateral cooperation in the hydrogen sector.

Overall, these innovations point to the emergence of a more comprehensive US green foreign energy policy that spans from climate and energy issues to trade, investments, and industry. They also illustrate the development of an explicit and coordinated green power strategy that relies on the deployment of diplomatic tools of soft power as well as on economic tools. As in the previous period, the economic tools sustain a catalytic power approach based on mobilization, public–private partnerships, and network diplomacy such as in the case of Power Africa, PGII, or the regional initiatives discussed above; however, they also impart an economic power perspective that is exemplified by the role that institutions such as the EXIM and the DFC play in promoting the internationalization of US green business and the diversification of low-carbon supply chains. Moreover, the shift towards a green economic power strategy under Biden is further visible in several IRA provisions that combine economic rewards and targeted regulatory measures. With these, Washington unilaterally leverages its market power not to promote rules but to increase national industrial capacity and restructure low-carbon supply chains according to its interests, particularly to reduce dependency on China (as explained, Biden also maintained the trade tariffs established by Trump on Chinese solar technology and EVs imports from China). This process entails an external dimension, which is another manifestation of US green economic power that is deployed bilaterally when Washington negotiates access to the IRA's green incentives.

Initially, the IRA incentives for EVs were designed for those countries with a free trade agreement in place with the US. These were twenty when the measure was enacted: Australia, Bahrain, Canada, Chile, Colombia, Costa Rica, Dominican Republic, El Salvador, Guatemala, Honduras, Israel, Jordan, South Korea, Mexico, Morocco, Nicaragua, Oman, Panama, Peru, and Singapore. However, in early 2023 the USDT issued guidelines specifying that this opportunity could be open to additional countries on the basis of a more specific sectoral agreement. In March 2023, Japan was the first country to negotiate a Critical Minerals Agreement with Washington to enable its industries and technologies to access the IRA incentives. As illustrated, in 2021 the US and the EU had established a Trade and Technology Council to increase transatlantic cooperation in the realm of green industrial policy. This was followed by the establishment, in 2022, of the US–EU Joint Task Force on the IRA, which was intended to address the EU's concerns about its negative impact on European industrial development. In the summer of 2023, then, the US and the EU decided to initiate negotiations on a Critical Minerals Agreement (still pending). Additional countries, such as the UK, expressed their intention to follow a similar approach. Moreover, in December 2023, the USDT and the DoE proposed more specific regulations for implementing FEOC's IRA provisions (which should be formally approved in early 2024). They confirmed that such restrictions would apply to China, Russia, North Korea, and Iran. Hence, these rules would mostly preclude incentives for EVs containing battery components or critical minerals produced in China. However, recognizing the global integration of the EV industry and supply chains, these rules would not exclude from the IRA incentives those EVs with components or critical minerals produced outside China by subsidiaries of Chinese companies—or those by joint ventures or collaborations with such companies—that are not owned or directly controlled by the Chinese government (Jack et al. 2023). A 25% ownership threshold for determining whether a company is controlled by a FEOC was also proposed by the Biden administration. Given China's dominant position in the global battery supply chain, often in the form of joint ventures with Western partners, these rules could have far-reaching consequences for many producers around the world (Home 2023).

Finally, the US green power strategy included a military component, specifically in the Indo-Pacific, where the Biden administration sought to shape the wider institutional landscape of the region through its economic and security partnerships (Figure 5.6). Indeed, US green engagements in the Indo-Pacific must be placed in the context of the US strategy to counter China's influence by building 'collective capacity with allies' (The White House 2022d, 5). This is illustrated by the Biden administration's upgrading of the Quad and by the launch, in 2021, of the new trilateral military partnership with Australia and the UK (i.e., the AUKUS), which involves the assistance of the US and the UK in Australia's acquisition of nuclear-powered submarines, along with cooperation in other security-related areas. This allows Australia to increase its (military) capability and contribute to the US's effort to contain China's maritime rise. The Quad had already been relaunched under Trump; Biden strengthened this partnership even more by

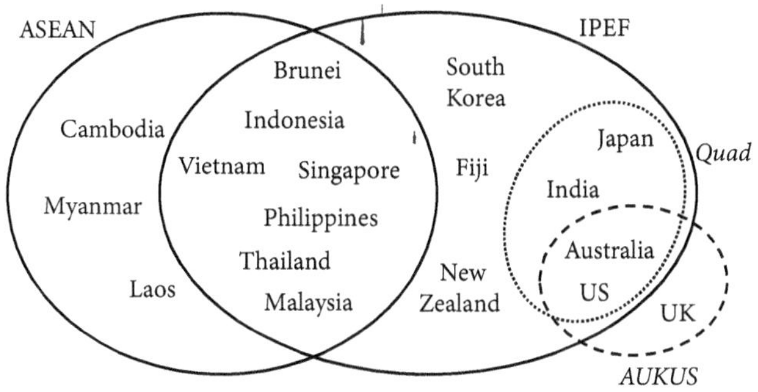

Figure 5.6 US Indo-Pacific strategy and the new regional institutional landscape

Note: Australia, Brunei, Cambodia, Indonesia, Japan, South Korea, Laos, Malaysia, Myanmar, New Zealand, the Philippines, Singapore, Thailand, and Vietnam are part (alongside China) of the RCEP

Source: Author's elaboration

directly adding a green component. In 2022 it promoted the Quad Climate Change Adaptation and Mitigation Package in order to enhance cooperation on climate action and the energy transition. This cooperation was strengthened in 2023 by a focus on low-carbon technology supply chains, technology standards, and clean hydrogen and ammonia. In 2023, the US also promoted a Quad Investors Network as a private sector-led platform to facilitate investments in strategic sectors, such as low-carbon technology and critical minerals.

Finally, this nexus of the US security and green agendas is visible in the case of Washington's engagements with the Philippines. In 2022 they reached a new security agreement, according to which US naval forces would gain access to four additional Philippine military bases. This major expansion to the 2014 US–Philippines Enhanced Defence Cooperation Agreement was facilitated by Manila's concerns about the increasingly assertive presence of China in the South China Sea. However, the Biden administration flanked this security partnership with the establishment of an Energy Policy Dialogue centred on renewable energy (as well as nuclear energy) and cooperation in the area of critical minerals supply chains. Towards this latter goal, Washington committed to support, through the USTDA, the development of a nickel- and cobalt-processing facility in the Philippines (The White House 2022f).

6

China, the European Union, and the United States

A Comparative Perspective

Comparing the green superpowers

The analytical framework developed in Chapter 2 has guided this empirical study of the green foreign energy policies and green power strategies of China, the EU, and the US. In this chapter, I offer a structured, focused comparison of these green superpowers based on that analytical framework. First, I reassess the historical trajectory and the current status of each of their green foreign energy policies and green power strategies, pointing to the key turning points and the transformations that occurred, from minor adjustments to major changes. Then, in the following section, I highlight their main differences and similarities. I explain these variations—in terms of issue areas, geographical reach, modes of engagement, tools mix, and overall green external approach—by examining the patterns of state–market relations in each case, their preferences and capabilities, and the internal and external environments. Finally, in the last section of this chapter, I discuss the effectiveness of the green power strategies of China, the EU, and the US in terms of limitations and potentials. These limitations and potentials consider both their capabilities and weaknesses and the constraints and opportunities offered by the changing international landscape.

The origins of the green foreign energy policies of China, the EU, and the US can be traced to the 1990s. All three took part in the multilateral process that resulted in the UNFCCC and, later, in the Kyoto Protocol. However, while the EU and the US both deployed their diplomatic tools and green soft power to shape the nascent global climate regime, Beijing was primarily a policy taker at that time. China's diplomatic engagement contributed to the Global South's positioning on the matter (e.g., with the CBDR principle), but it can hardly be considered a manifestation of a green power strategy. On the contrary, Beijing's approach was instrumental in preserving the role of fossil

Green Superpowers. Andrea Prontera, Oxford University Press. © Andrea Prontera (2024).
DOI: 10.1093/9780191987304.003.0006

fuels as a key ingredient of the developing world's growth strategies. Although the US position on the matter was similarly conservative, Washington was able to promote several of its preferences in the multilateral climate regime, as with the market-based instruments included in the Kyoto Protocol (regardless of the fact that the US, in the end, did not join it). The EU adopted a more committed position on climate change and multilateral efforts to limit emissions and promote mitigation measures. Its green soft power strategy was based on diplomatic engagement, coupled with efforts to lead by example.

In sum, the first manifestations of the US and EU green foreign energy policies focused on the climate issue area and diplomatic multilateral engagements, and they were based on the use of soft power resources to shape the global transition from fossil fuels to renewables (Table 6.1). It is worth noting that in the 1990s, with the NAFTA, the US created a (regional) regulatory regime for cooperation and trade in the area of electricity and renewable energy. This, however, did not represent an explicit manifestation of a green regulatory power strategy, partially due to the poor development of domestic renewable energy policy.

During the 2000s, the EU honed its green soft power. The focus on the UN multilateral climate regime continued, and its leadership by example was strengthened, including by the launch of the ETS in 2003. However, since the mid-2000s, the European Commission has promoted several diplomatic bilateral initiatives that span from the climate to the energy issue areas and target several countries around the world. Moreover, the EU began to deploy regulatory tools in its green foreign engagements following parallel developments at the domestic level, although the results of this innovation, which was aimed at exporting its rules and preferred policies, became evident only in subsequent years. The US approach also changed in the 2000s. Under the George W. Bush presidency, the US disengaged from the global climate regime. Its green foreign energy policy shifted from multilateral modes of engagement towards minilateral and bilateral modes and into the energy issue area. Still, the US green power strategy was centred on soft power resources, channelled on a global scale through a set of new institutions for increasing cooperation and dialogue on renewables and low-carbon technology, which were used to co-opt both major industrialized countries and emerging economies and to promote Washington's views on the global energy transition.

China's green foreign energy policy emerged during this period. It was very different in terms of issue areas and tools from those of the EU and the US (Table 6.1). It was embedded in the Beijing 'going out' policy and focused on the trade and investment issue area, specifically concerning the mineral

Table 6.1 Comparing the green superpowers (I): Chinese, EU, and US green foreign energy policy (GFEP) and green power strategy (GPS) (1990–2010)

		1990s	2000s
China	GFEP	*Tools*: diplomatic *Modes*: multilateral *Issue areas*: climate *Reach*: global	*Tools*: diplomatic, economic *Modes*: multilateral, bilateral *Issue areas*: climate, energy, and trade and investment *Reach*: global
	GPS	*Not developed.* China participates in the UNFCCC process, but it is a policy taker with limited interests and capabilities to shape and promote the energy transition abroad	*Economic power.* State-owned banks and commercial diplomacy mobilized (in the context of the 'going out' policy) to sustain the internationalization of national companies, enhance trade links in the mineral sector, and structure low-carbon supply chains on a regional and global scale *Soft power.* Increased bilateral engagements in the area of renewable energy (especially towards the Global North); China is still a policy taker in this field as well as in the area of climate change
EU	GFEP	*Tools*: diplomatic *Modes*: multilateral *Issue areas*: climate *Reach*: global	*Tools*: diplomatic, regulatory *Modes*: multilateral, minilateral, bilateral *Issue areas*: climate, energy *Reach*: global (diplomatic tools), regional (regulatory tools)
	GPS	*Soft power.* Diplomacy, and leadership by example with committed and ambitious position on climate mitigation measures and multilateral efforts	*Soft power.* Diplomacy, ideas, and (market-based) policies, leadership by example with committed and ambitious position on climate mitigation and multilateral efforts. Bilateral engagements in the area of climate change and renewable energy in order to reinforce the global climate regime and promote and shape the energy transition abroad *Regulatory power.* International institutions and regulatory regimes (e.g., EEA, EnC) to export the EU rules at regional level through bilateral and minilateral engagements
US	GFEP	*Tools*: diplomatic *Modes*: multilateral *Issue areas*: climate *Reach*: global *Note*: Regulatory tools and regional reach for the NAFTA	*Tools*: diplomatic *Modes*: minilateral, bilateral *Issue areas*: climate, energy *Reach*: global

	1990s	2000s
GPS	*Soft power*. Diplomacy, ideas, and (market-based) polices for shaping the global climate regime. Less committed position on climate mitigation measures and multilateral efforts *Regulatory power*. NAFTA's regulatory regime for electricity and renewable energy trade (poorly developed green power strategy)	*Soft power*. Disengagement from the global climate regime, but diplomacy, ideas, and (market-based) polices, and institution-building—new venues for communication and policy exchanges—for co-opting other actors and shaping the energy transition through (mainly) minilateral and bilateral modes of engagement

Source: Author's elaboration

sector. Within this sector, China deployed its economic tools and commercial diplomacy to sustain the internationalization of its companies and create trade links that were instrumental in tapping the growing domestic demand for its expanding green manufacturing capacity. These foreign engagements, coordinated with domestic industrial policies, helped in the structuration of Chinese-centred low-carbon supply chains on regional and global scales. During the 2000s, China also increased its diplomatic engagement in the area of climate change and (chiefly) renewable energy technology. However, the majority of Beijing's (bilateral) agreements on the matter involved countries in the Global North—including the US, the EU, and European green pioneers—in relation to which China was a policy taker. China was also a policy taker with regard to the innovations introduced in the Kyoto Protocol, such as the CDM initiative. This Chinese green power strategy continued into the early 2010s (Table 6.2). However, since this period, China has deployed its economic tools and commercial diplomacy also to support the internationalization of its green business. Moreover, with the arrival of Xi Jinping and the launch of the BRI, Beijing further escalated its global investment and trade agenda through the realization of new infrastructure and energy projects. Like the 'going out' policy, the BRI was initially centred on fossil fuels. Later, it gradually incorporated a green dimension, which manifested through the growing involvement of Chinese companies (especially SOEs) and state-owned banks in the realization of wind and solar projects abroad. These engagements were not limited to the BRI partners, although the Chinese green approach took a more strategic stance towards these countries. In this case, Chinese green economic power was coordinated with the wider Chinese approach to extend its influence in key regions of the Global South. With the Paris Agreement, China decided to play a more constructive role in the multilateral climate regime—a move that was paralleled by an increasing focus on climate change in several minilateral (regional) forums established by Beijing. These helped China deploy its green soft power and foster its view on climate action and energy transition, which includes a strong role for the state and state-led developmental approaches. Conversely, China has not deployed regulatory tools, sought to build regulatory regimes, or followed a 'regulation through trade' logic to promote rules or policy changes abroad. Even to promote transnational electricity networks or the diffusion of its technical standards, China has mainly relied on the foreign activities of its companies.

In contrast, the EU and US green power strategies have changed significantly since the end of the 2000s (Table 6.2). In both cases, the 2008 economic crisis triggered important transformations in state–market relations with significant external repercussions. In addition, both green foreign energy policies evolved in response to China's rise in terms of industrial capacity and the external influence acquired with the BRI. In the US, the election of Obama was a decisive step in this major shift. Under Obama, the US re-engaged with the global multilateral climate regime. The country's green soft power continued to be channelled through minilateral and bilateral initiatives, but these were now intended to complement rather than challenge the UN climate efforts. To reinforce its global green posture, the US also increased funding for multilateral initiatives and sought to redirect Western-led multilateral development banks' actions towards the needs of climate mitigation and the energy transition. Moreover, under Obama, the US attempted to expand its use of economic and regulatory tools, the latter of which had been absent from the US approach, with the exception of the NAFTA. In addition, Washington did not use its trade deals according to a 'regulation through trade' logic. At the same time, Obama, as part of his wider liberal international agenda, launched the TTIP and TTP, which were designed to facilitate renewable energy trade as well as offer—in the Indo-Pacific—an alternative model to the Chinese approach based on SOEs and BRI engagements. These regional initiatives were eventually abandoned. However, new financial resources were channelled by the US development and aid agencies to promote and shape the energy transition in several countries of the Global South. US aid agencies had already been involved in promoting small renewable energy projects in developing countries; now their engagements

Table 6.2 Comparing the green superpowers (II): Chinese, EU, and US green foreign energy policies (GFEP) and green power strategies (GPS) (2010–2024)

		2010s	2020–2024
China	GFEP	*Tools*: diplomatic, economic *Modes*: multilateral, minilateral, bilateral *Issue areas*: climate, energy, industry, trade and investment *Reach*: global	*Tools*: diplomatic, economic, regulatory *Modes*: multilateral, minilateral, bilateral, unilateral *Issue areas*: climate, energy, industry, trade and investment *Reach*: global, regional (for regulatory regimes)
	GPS	*Economic power*. SOEs, state-owned banks, and commercial diplomacy mobilized (in the context of the BRI) to sustain the internationalization of national green business, promote renewable energy projects abroad, and enhance investments and trade links in the mineral sector *Soft power*. After sustaining the multilateral climate efforts that led to the Paris Agreement, China channelled soft power resources for shaping the energy transition especially through minilateral forums and the BRI initiatives (Chinese state-led green developmental approach as an example for countries in the Global South)	*Economic power*. SOEs, state-owned banks, and commercial diplomacy to sustain the internationalization of national green business (and Chinese green technologies and standards) enhance investments, industrial partnerships and trade links in the low-carbon technology and mineral sectors (further 'greening' of the BRI, and green foreign objectives incorporated into the wider Chinese global strategy). State control extended over the domestic minerals sector and key green technologies (potentially to be used as a 'stick' vis-à-vis other states) *Soft power*. After the 2020 climate neutrality pledges, China continued to channelled soft power resources for promoting and shaping the energy transition through minilateral forums, the BRI initiatives, and an increasing role for climate finance. However, Beijing did not embrace the role of a climate leader *Note*: Upgrading of unilateral measures for 'greening' the BRI and Chinese foreign investments. RCEP as a regulatory regime for increasing renewable energy trade and technical cooperation in the Indo-Pacific (but not a developed green regulatory power strategy)
EU	GFEP	*Tools*: diplomatic, regulatory, economic *Modes*: multilateral, minilateral bilateral *Issue areas*: climate, energy, trade and investment *Reach*: global, regional (for regulatory and economic tools)	*Tools*: diplomatic, regulatory, economic *Modes*: multilateral, minilateral, bilateral, unilateral *Issue areas*: climate, energy, industry, trade and investment *Reach*: global

Continued

Table 6.2 *Continued*

	2010s	2020–2024
GPS	*Soft power.* Diplomacy, ideas, and (market-based) policies, leadership by example. Bilateral and minilateral engagements in the area of climate change and renewable energy for promoting and shaping the energy transition abroad and reinforcing the global climate regime. 'Sustainable development' clauses in trade deals *Regulatory power.* Upgrading and greening of the EU regulatory regimes (e.g., EEA, EnC) to export the EU rules at regional level through bilateral and minilateral engagements *Catalytic power.* Expanded use of economic tools in a catalytic power fashion (e.g., blending facilities and guarantees) for promoting renewable energy projects abroad (with a focus on the EU neighbourhood and Africa)	*Soft power.* Diplomacy, ideas, and (market-based) policies, leadership by example. Bilateral and minilateral engagements in the area of climate change and renewable energy for promoting and shaping the energy transition abroad and reinforcing the global climate regime. 'Sustainable development' clauses in trade deals. Bilateral and minilateral engagements leveraged also to increase cooperation in the area of industry, low-carbon supply chains, standards and regulatory matters with 'like-minded' countries *Regulatory power.* Further upgrading and greening of the EU regulatory regimes to export the EU rules at regional level through bilateral and minilateral engagements. Sharpening of the 'regulation through trade' approach. Trade deals used to increase the resilience of the EU low-carbon supply chains and reduce dependency on China. Unilateral regulatory measure with potentially a global reach (e.g., CBMA) *Catalytic power.* Expanded use of economic tools in a catalytic power fashion (e.g., EFSD+) with a focus (mainly) on the EU neighbourhood and Africa (but global reach for the JETPs). Better coordination with the other means of the EU green external action, and expansion towards the industry issue area and new goals of increasing the resilience of the EU low-carbon supply chains and reduce dependency on China *Note*: Increasing efforts to develop a green economic power strategy (especially since 2022–2023)

US	GFEP	*Tools*: diplomatic, economic, regulatory *Modes*: multilateral, minilateral, bilateral, unilateral *Issue areas*: climate, energy, industry, trade and investment *Reach*: global	*Tools*: diplomatic, economic, regulatory, military *Modes*: multilateral, minilateral, bilateral, unilateral *Issue areas*: climate, energy, industry, trade and investment *Reach*: global
	GPS	*Soft power*. Relaunch of the US green soft power (e.g., US joining the Paris Agreement): multilateral, minilateral, and bilateral diplomacy to sustain global climate efforts and the energy transition. Ideas and (market-based) polices, and institution-building for promoting and shaping the energy transition. Increasing funding channelled through the US development and aid agencies for pursuing these objectives in the Global South *Catalytic power*. Expanded use of economic tools in a catalytic power fashion for promoting renewable energy projects abroad (e.g., Power Africa) *Economic power*. State-owned banks and commercial diplomacy to support the internationalization of national green business, which is also protected with tariffs and trade measures *Note*: Under Trump, reduction of the US diplomatic and economic engagements on climate change and renewable energy and overall diminishing of US green soft power, catalytic power, and economic power (although implementation of trade tariffs on Chinese solar technology and EV imports from China). In the critical mineral sector, diplomacy (minilateral and bilateral) and initial use of economic tools to increase supply chain resilience. Transformation of the NAFTA into the USMCA (regional regulatory regime for renewable energy and low-carbon technology)	*Soft power*. Relaunch of the US green soft power (e.g., US rejoining the Paris Agreement); multilateral, minilateral, and bilateral diplomacy to sustain global climate efforts and the energy transition. Ideas and (market-based) polices, and institution-building for promoting and shaping the energy transition. Increasing funding channelled through the US development and aid agencies for pursuing these objectives in the Global South. Bilateral and minilateral engagements leveraged also to increase cooperation in the area of industry, low-carbon supply chains, standards and regulatory matters with allies and 'like-minded' countries *Economic power*. State-owned banks and commercial diplomacy to support the internationalization of national green business (which is also protected with tariffs and trade measures) and increase the resilience of low-carbon supply chains. Unilateral regulatory measures and bilateral trade deals for restructuring low-carbon supply chains and reducing dependency on China. Military alliances and partnerships linked to green foreign energy policy objectives (especially in the Indo-Pacific) *Catalytic power*. Expanded use of economic tools in a catalytic power fashion. Expansion towards the industry issue area and new goals of increasing the resilience of the US low-carbon supply chains and reduce dependency on China

Source : Author's elaboration

significantly intensified as part of a strategy based on a combination of soft power and catalytic power resources, such as with the Power Africa initiative. However, the US also embraced a green economic power approach under Obama, while Washington's green foreign energy policy expanded towards the industry and trade and investment issue areas.

This major change represented the external dimension of the upgrading of the US green developmental state approach and was aimed at supporting national green business. This goal was pursued by leveraging the armamentarium of US foreign economic policy and commercial diplomacy (e.g., the OPIC and EXIM), as well as by deploying other traditional tools of statecraft, such as tariff and trade measures, both within the context of the WTO and unilaterally, such as in the case of the dispute on solar technology with China. With the arrival of Trump, the US disengaged from the multilateral climate regime, a move that hindered US green soft power. Washington did not develop an explicit and articulated green foreign energy policy or green power strategy under Trump. It reduced its diplomatic and economic engagements to promote and shape the energy transition and support national green business abroad. However, some innovations introduced under Obama persisted, and the US green soft power and catalytic power approaches were not completely dismantled. With the USMCA replacing NAFTA, the US also maintained a regional framework for regulatory cooperation and trade with relevance for renewable energy and low-carbon technology. Moreover, Trump continued with trade tariffs targeting Chinese solar technology and established new ones on EV imports from China. Finally, under Trump, the US green foreign energy policy increased its focus on reducing vulnerabilities along low-carbon supply chains and promoting bilateral and minilateral cooperation with allies and potential partners. This was the result of growing rivalry with China, which also manifested in a relaunch of the US economic tools towards the end of the Trump administration. Again, however, this process was not incorporated into a well-defined green power strategy.

The 2010s was also a period of major change for the EU (Table 6.2). Unlike in the US, these transformations followed an incremental path throughout the second Barroso Commission and the Junker Commission. In this period, the greening of the EU regulatory state was paralleled with the growing use of regulatory tools to promote and shape the energy transition abroad. The EU sought to export its preferred rules and views through various channels. At the regional level, it embedded its green turn into the policy and regulatory frameworks it had built for managing relations with its neighbours. Only in limited subregions, however, was the EU actually able to deploy its green regularity power—e.g., in the Western Balkans, those countries that joined the EnC, and the EEA countries—whereas in other cases, the EU strategy remained primarily reliant on soft power. The EU continued to wield its green soft power in the multilateral climate regime as well as through bilateral engagements on energy and/or climate change on a global scale, both of which also served to reinforce the EU's influence beyond a 'leadership by example' role. Bilaterally, the EU also tried to promote the energy transition by including 'sustainable development' clauses in its trade deals. This deployment of a 'regulation through trade' logic on a global scale, however, was (in practice) still in line with a green soft power strategy.

These innovations illustrate the expansion of the EU green foreign energy policy towards the trade and investment issue area as well as an increasing intensity in the use of the diplomatic and regulatory tools that were already part of the EU's arsenal. More important changes resulted from the 2008 economic crisis and the evolving international landscape, including the ascendency of China. These events favoured a gradual shift in state–market relations in Europe and the emergence of a catalytic state in the EU green realm. This form of state represented the EU version of an emerging developmental state approach based (also) on the expanded use of economic tools and increased involvement by the EU development banks in the area of low-carbon technology, renewable energy, and energy networks. In turn, this evolution helped to construct a new external reach for the EU, with the parallel rise of a green catalytic power approach in its foreign engagements. This aimed to promote the energy transition abroad by supporting investment projects at regional and global levels (primarily the former). This was (in part) a continuation of the EU development policy's focus on sustainable development objectives, as well as of the foreign engagements of the EIB on the matter. Yet, since the mid-2010s, economic tools, such as blending facilities and guarantees, have shifted from a niche in the EU development policy to the broader machinery of the bloc's foreign policy, which after the Paris Agreement and the Energy Union Strategy better incorporated the energy transition's objectives. This change, however, did not push the EU towards a green economic power strategy; commercial diplomacy remained outside the scope of the EU foreign engagements, the bloc avoided escalating trade disputes on renewable energy technology, and it confirmed its preferences for a rule-based order centred on the WTO. It is also worth noting that, as explained, the EU catalytic power approach differs from that of the US. Indeed, for the EU this is a way not only to create coalitions with foreign actors, but also to increase coordination among the European actors themselves, i.e., the member states and their development banks, and to improve the overall role of the European Commission in the bloc's green external relations.

Another critical juncture for the green superpowers occurred between 2019 and 2022, when the effects of the Covid-19 crisis and of the Russian invasion of Ukraine reinforced each other. This combination fostered a certain convergence among the green foreign energy policies and green power strategies of China, the EU, and the US (Table 6.2). At home, all three actors adopted a developmental state approach increasingly imbued with strategic elements, further shifting towards forms of 'domestically oriented' statecraft in the area of renewables and low-carbon technology supply chains. Internationally, they all made efforts to formulate comprehensive approaches—in terms of issue areas, tools, and goals—to the external dimension of the energy transition, which had been fully integrated into their broader strategy to cope with a more challenging external environment.

China continued to advance its green economic power approach, including through the further greening of the BRI. This also served to expand Beijing's green soft power, which was improved by the ongoing evolution of Beijing's position on climate policy at home and abroad. China did not embrace the role of a climate leader, and its position on fossil fuels, including coal, continued to be conservative and cautious. However, its post-2020 climate neutrality pledges, coupled with its enduring (state-led) green industrial and technological rise, offer an interesting example to several countries of the Global South. During this period, China also increased its funding for capacity building and renewable energy projects in developing countries. Moreover, it supported the expansion of climate financing in non-Western-led multilateral development banks, such as the NDB and the AIIB. At the same time, China incorporated green foreign energy policy objectives into its strategy to extend its influence in the international system, challenging US (and Western) leadership at the global and regional levels. Renewables and low-carbon technology became part of Beijing's overall external relations, from minilateral forums like the BRICs to regional security partnerships, such as the SCO, to the Chinese economic and diplomatic engagements, which spanned from Asia and the Middle East to Africa and Latin America. As in previous periods, however, bilateral ties, commercial diplomacy, and Chinese companies' foreign investments remained the main components of Beijing's green (economic) power strategy, which maintained a mercantilist edge and aimed primarily at promoting national green business (including expansion into new sectors, such as EVs and hydrogen) and addressing possible vulnerabilities along critical minerals supply chains that could affect the growth of domestic industrial capacity and the energy transition at home.

To achieve this latter objective, Beijing also resorted to traditional means, such as extending state control over the national mineral sector or issuing regulations to potentially ban the export of key PV technologies. This sharpening of Beijing's armamentarium for green economic statecraft might help China in case of trade confrontations with the West. Since 2020, China has also sought to intensify the use of regulatory tools. This is not according to a 'regulation through trade' logic but to set unilateral measures in order to push China's investments abroad towards higher green standards. This move might turn China into a rule exporter through its overseas investments, although the actual impact is still unclear. Moreover, by backing the RCEP, China has supported the creation of a regulatory regime that can favour regional trade in renewable energy technology and improve technical cooperation in the Indo-Pacific region. This move, however, has not facilitated the creation of a clear and well-developed green regulatory power strategy by Beijing. Finally, in order to become a global standard setter, China has also increased its engagement with international standards development organizations. This represents a (step towards) change with respect to its traditional approach, according to which the diffusion of Chinese technological standards was pursued primarily through the foreign investments of Chinese companies and through the BRI's network of commercial and trade agreements.

As with the 2008 economic crisis, the Covid-19 crisis triggered changes in state–market relations that affected the US and EU green approaches at home and abroad (Table 6.2). This trend was then amplified by the war in Ukraine and the growing rivalry between the US and China. The election of Biden was another key change in the US. His administration set the US back on the green path initiated by Obama, although with a sharper approach vis-à-vis China. It formulated a comprehensive green foreign energy policy and better linked internal and external measures for deploying the country's green power. Under Biden, the US re-engaged with the multilateral climate regime and relaunched its green soft power to promote and shape the global energy transition through development aid and diplomatic engagements, as well as by establishing new international initiatives for both industrialized countries and emerging economies. The Biden administration also intensified the use of economic tools in the US green foreign energy policy, again following both green economic power and green catalytic power approaches in order to promote the energy transition abroad, reduce vulnerabilities along low-carbon supply chains, and support national green business. An additional aspect of the US green catalytic power was a focus on countering China's influence in various regions of the world by creating coalitions of allies and partners as well as by mobilizing private financial resources. Moreover, unlike Obama, Biden did not try to relaunch an international liberal agenda or revitalize the WTO. His administration emphasized a more mercantilist approach based on strategic deployment of the country's economic might. This manifested through a combination of internal and external measures, including trade deals aimed at supporting and expanding the US green industrial capacity, restructuring low-carbon supply chains—from critical minerals to manufacturing—to prioritize Washington's allies and partners, and reducing dependency on China. Protectionist (unilateral) measures, especially those included in the IRA, created friction with the US's allies. However, under Biden, the US sought to reinforce green cooperation with them through bilateral and minilateral engagements. In so doing, the US connected its military posture and evolving security partnerships with green foreign energy policy objectives, especially in the Indo-Pacific region, where the US's commercial diplomacy and military footprint reinforced each other.

With the launch of the EGD by the von der Leyen Commission, alongside the innovations introduced during the Covid-19 crisis and the war in Ukraine, the EU also formulated a comprehensive green foreign energy policy. This was mostly based on upgrading and better coordinating the existing tools and refocusing towards the industry issue area and the goal of reducing vulnerabilities along low-carbon supply chains (from manufacturing to critical minerals). Additionally, the energy transition was fully embedded in all the external diplomatic and economic engagements of the EU. Support for the multilateral climate regime was confirmed, as was the EU's global green soft power—although the EU increased its focus on bilateralism and minilateralism with 'like-minded' countries. The externalization of the EU green regulatory state in its neighbourhood continued, and the 'regulation through trade' logic was sharpened. Moreover, the EU increased efforts to (unilaterally) deploy its green regulatory power, such as with the regulations on batteries and green hydrogen that aim at setting global standards aligned with the EU preferences in these fields. This approach was elevated with the CBAM, which can potentially augment the global reach of the bloc's green regulatory power. At the same time, the EU began to use trade deals to increase the resilience of its low-carbon supply chains. Financial support for promoting renewable energy abroad and increasing the resilience of the EU low-carbon technology supply chains also increased, following a green catalytic power perspective (e.g., with the EFSD+), which now also served to support the EU technological standards abroad and to sustain the bloc's geopolitical turn, both of which were enshrined in the Global Gateway Strategy. The financial resources of the EU continued to be deployed mainly in its neighbourhood. However, with the JETPs, which were created in cooperation with the US and other industrialized countries, the EU's influenced swelled to a global scale. Moreover, to fill domestic gaps in the area of renewables and green hydrogen, the EU deployed its green foreign energy policy tools to promote transnational electricity networks and hydrogen corridors that involve neighbourhood countries. The EU also sought to move towards a green economic power approach to uphold its geopolitical turn, a transition that was accelerated by the war in Ukraine. To accomplish this, the EU improved its 'defensive' regulatory armamentarium with tools such as the Anti-Coercion Instrument and a new framework for investment screening. In the area of green hydrogen, the EU established a common (voluntary) platform (i.e., the EU Energy Platform) for pooling demand and leveraging bloc market power. The CRMA also provided for a joint purchasing mechanism to aggregate demand for critical minerals at the European level. Both the CRMA and the NZIA also include regulations and unilateral measures to reduce dependency on China in several critical low-carbon supply chains. The European Commission even advanced path-breaking proposals, such as an EU sovereignty fund or an EU export credit agency, to support the internationalization of European green business. These proposals, however, were not backed by member states. The EU's efforts to increase European green industrial capacity and the resilience of its low-carbon supply chains hence continued to follow a (mostly) catalytic approach at home and abroad. However, some signs of a (possible) nascent green economic power strategy, albeit not yet fully implemented, began to appear.

Differences and similarities in green foreign energy policy

Over time, the green foreign energy policies of China, the EU, and the US have, to a certain extent, converged. The three green superpowers have extended the scope of their actions to encompass all the issue areas of this composed policy field, intensified the use of all the tools at their disposal to achieve the policy's main goals, and engaged in different modes in several countries in their region and on a global scale. Moreover, all three green superpowers have fully embedded green foreign energy policy objectives into their broader, evolving international strategy. Yet, important differences still persist among them, as the intensity and emphasis of their actions vary in terms of issue areas, geographical reach, modes of engagement, tools mix, and overall external green approach. As China, the EU, and the US present very distinct systems of political economy, diverse institutional settings, and specific trajectories of economic development, their foreign relations obviously reflect these differences. More specifically, however, the main variances in their green foreign energy policies can be explained by considering the evolving patterns of state–market relations in each case, their preferences and capabilities, and the internal and external environments in which they operate.

Since its inception, China's green foreign energy policy has been closely linked to its developmental state approach, which assumes a strong role for the state in energy and industrial governance that includes protecting and nurturing national companies in strategic sectors. This state–industrial complex is at the core of Beijing's green approach at home and abroad, as well illustrated by the prominent role that SOEs and state-owned banks play in the domestic renewables and low-carbon sectors and in supporting the internationalization of Chinese green business, technologies, and standards. The main emphasis of China's green foreign energy policy, hence, is on industry, trade and investment, commercial diplomacy, and bilateral modes of engagement. These serve to reinforce the external reach of the Chinese green developmental state and sustain the energy transition at home. In turn, the export of Chinese green technology, SOEs' investments, and state-backed financial assistance help Beijing to strengthen relations with third countries, especially in the Global South. This approach resonates with China's preference for a 'hub and spokes' model of international (economic) relations (e.g., Shaffer and Gao 2020), a model that simultaneously shaped the BRI and was reinforced by it. This model depends on Chinese industrial capacity and on the ability of the state to directly mobilize economic and financial resources as well as leverage access to its vast market in order to attract other countries to its web of influence. It is additionally influenced by domestic constraints, as the legitimacy of the CCP is linked to the ability of its leadership to maintain sustained economic and industrial growth. Unlike its Western counterparts, capitalism's boom and bust cycles do not pose a challenge just for the Chinese government, but for the party-state centred political system itself.

These features were and are the core of the Chinese green foreign energy policy and green power strategy. Innovations, however, have appeared, especially in the last decade. Since the arrival of Xi Jinping, the control of the party-state over the market and key sectors of the national economy has improved. Moreover, the Chinese green developmental state has acquired a more strategic dimension as renewables and low-carbon technology have become a new field of competition among the great powers and as the overall Chinese international posture has become more assertive, triggering Western responses that have worsened the external environment for Beijing's green rise. At the same time, since the mid-2010s, China has (gradually) come to play a more constructive role in the multilateral climate regime. The legitimacy of the CCP is also linked to addressing environmental degradation and the negative consequences of the climate crisis at home. Tackling pollution was initially an important driver of China's renewable energy policy. However, this can hardly be considered a constraint on Beijing's external action, although the call for 'greening' the BRI was also sustained by domestic actors. By joining the Paris Agreement and by pledging carbon neutrality targets, China aligned its national green trajectory to global climate efforts and showed its willingness to act as a responsible great power. However, it has not drastically changed its position on fossil fuels, including coal, which is still seen as an essential component of the country's energy security and economic development. Hence, China has continued to advocate a cautious approach to the energy transition and to ask for more decisive actions, especially from the Global North. This position confirms the longstanding preferences of Chinese foreign policy, which is guided by principles of non-interference and sovereignty, with multilateral interactions serving primarily domestic policy goals (e.g., Qin 2014; Sutter 2016). As China has become the major producer of several green technologies, the global energy transition offers Beijing additional opportunities for industrial development and augmenting its international influence. Yet, through the recent greening of the BRI and due to several minilateral (regional) forums and bilateral dialogues, as well as to the efforts to increase (bilateral and multilateral) climate finance, China has also activated additional channels for deploying soft power resources. Its preference for non-interference, however, limits its willingness to promote specific policies or reforms for supporting and shaping the energy transition in developing regions, although its state-led green industrial model can offer a convenient example for several of those countries. This preference for non-interference also limits the use of regulatory tools in Chinese green foreign energy policy, including the use of the 'regulation through trade' logic. Unilateral regulatory measures for greening its overseas financial engagements have intensified since 2020. However, it is too soon to understand the practical impact of this Chinese green regulatory apparatus, which is still weak vis-à-vis the country's developmental state apparatus.

The EU green foreign energy policy differs widely from China's, reflecting the diverse nature of these two actors as well as different capabilities, preferences, and constraints. Some similarities have recently appeared in terms of objectives, but their tools and overall approaches remain very distinct. Initially, the EU green foreign energy policy targeted the multilateral climate regime and relied on diplomacy and a 'leadership by example' role with strong commitments to the UN processes and ambitious targets. This choice reflected the bloc's preferences for multilateralism, for a ruled-based international order—within which international commitments are not perceived as a limit to sovereignty—and for tackling climate change with incisive actions. However, this was also the only approach that the EU could follow, as its capabilities for green external action were very limited in the 1990s. During the 2000s, these capabilities gradually increased, and the EU intensified the use of diplomatic tools to promote and shape the energy transition through multilateral as well as bilateral modes of engagement. Moreover, during this period, the greening of the EU regulatory state began to have external repercussions. The EU deployed regulatory tools at the regional level, and in the context of the nascent ENP, in order to promote abroad its market approach and rules for energy governance, which gradually incorporated the internal move towards renewables. This development reflected the EU's capabilities and preferences as a regulatory state and resonated with the bloc's view on regional integration through regulatory regimes. Yet, even in its neighbourhood, the EU encountered several constraints, as many countries, especially in the Southern Mediterranean, were not eager to enter into legally binding agreements. Hence, the EU formulated a more flexible approach that combined softer means of conditionality—e.g., forums for policy exchanges, technical assistance, and transgovernmental networks—and more intrusive international institutions, i.e., legally binding treaties such as the EnC or the EEA. Through this complex (regional) system, the EU aimed to favour the diffusion of renewable energy (or the realization of electricity networks) in an indirect way by shaping the regulatory and policy framework of third states.

In the early 2010s, the EU green foreign energy policy continued to evolve. After the Lisbon Treaty, the EU had increased its competences in the area of energy policy and its capabilities for external action. At the same time, the integration of energy and climate policies had been improved since the Climate and Energy Package was issued in 2008. The external repercussions of the greening of the EU regulatory state augmented as well as the EU's diplomatic green efforts. The EU preferences continued to support multilateralism, but the EU also developed a more flexible strategy to cope with a changing international environment. Moreover, it embraced a 'regulation through trade' approach on a global scale. This primarily served as a vehicle for green soft power, as the bloc had limited resources to push for more incisive action vis-à-vis other countries in this regard. More importantly, it was during this period that the EU developed its green catalytic power approach, which resulted from the shift in state–market relations following the 2008 economic crisis. This EU-level version of a developmental state was further upgraded with the appointment of the Juncker Commission; in light of the challenging international landscape, the new EU green interventionist stance acquired a more strategic dimension. As discussed, the EU expanded its use of economic tools in foreign engagements. However, the EU had limited capabilities to go beyond green catalytic power and enter into the reign of green economic power. Nevertheless, the emergence of a green catalytic power strategy signified an important change: the bloc could now rely on more direct means to promote renewable energy projects and the energy transitions in third states, although this was still dependent on its ability to create multi-actor coalitions. These new capabilities—well illustrated by the EU blending facilities and the increased role of the EIB in the landscape of European development policy—were mainly directed towards the EU neighbourhood, reflecting the wider orientation of the bloc's external relations as well as its limited resources on this front.

Internal political constraints to climate action increased in Europe in the late 2010s. However, this did not halt the EU green turn, which received a further push under the von der Leyen Commission and its EGD agenda. In the wake of the Covid-19 crisis and the war in Ukraine, this agenda was even strengthened, resulting in the supplementary greening of the EU regulatory state and in an expansion of the EU catalytic state with an explicit focus on renewables and low-carbon technology. Due to the ongoing deterioration of the international environment, these efforts further shifted the EU towards forms of 'domestically oriented' statecraft aimed at protecting European green industry, increasing industrial capacity in strategic sectors, and reducing dependencies in critical low-carbon supply chains, particularly vis-à-vis China. This was matched by an evolution of the EU green catalytic power—e.g., with the Global Gateway and the EFSD+—and of the EU green soft power and regulatory power approaches, primarily due to the inclusion of green foreign energy policy objectives in all the components of EU external relations. Nevertheless, the innovations introduced in this period fell short of bringing the EU on the side of green economic power. Such an evolution was not supported by the member states. These internal constraints continued to limit the EU green foreign energy policy, even after the outbreak of the war in Ukraine. Moreover, despite the growing geopolitical tensions spurred by the war, the EU has not abandoned its preferences for multilateralism in international economic and trade relations, which contrast with more protectionist and mercantilist measures. This position reflects also inherent weaknesses of the bloc, as it needs cooperation to achieve its energy transition objectives and cannot adopt a more muscular and confrontational approach.

At the same time, the evolving international landscape has prompted the EU to increase emphasis on bilateralism, minilateralism—with 'like-minded' countries—and even unilateralism, such as with the CBAM. Even in the latter case, however, the EU has sought to deploy its green regulatory power by designing a measure that is compatible with the WTO provisions.

Like the EU, Washington initially targeted the multilateral climate regime and deployed its diplomatic tools to promote its views on the matter. These, however, were more conservative, mirroring US reluctance to support binding international commitments that could be perceived as limits to its national sovereignty. In addition, the important role of fossil fuel interests in domestic politics was a strong constraint on more ambitious mitigation policies. These preferences and constraints influenced the US green foreign energy policy throughout the 1990s, even under Clinton, when major efforts were made to constructively engage with the multilateral climate regime. The US was able to shape this regime by promoting its market-based preferences for climate governance, but Washington's green soft power was then hindered by its inability to sign the Kyoto Protocol. Domestic constraints, including the role of Congress vis-à-vis the president, and the growing polarization of climate issues ultimately hindered both Clinton administrations on this front. On the other hand, it was during this period, with the NAFTA, that the US created a regulatory regime for energy trade, market integration, and regulatory cooperation in North America. As explained, this was not a manifestation of a green regulatory power strategy, although the agreement reflected the US preferences for an international liberal agenda that was pursued globally with the WTO.

During the 2000s, under George W. Bush, the US disengaged from the multilateral climate regime. However, his administration established a set of minilateral institutional venues and bilateral dialogues on climate change, renewables, and low-carbon technology. These were sustained by the US diplomatic capabilities and the global reach of Washington's foreign policy and were instrumental for deploying soft power and promoting US preferences abroad, which were based on the idea that climate mitigation objectives would be better achieved through technological advances and market tools rather than binding commitments and stricter regulations on fossil fuel usage. This soft power strategy also entailed efforts to promote the global diffusion of market reforms for the energy sectors, including by providing capacity building and technical assistance. However, in this period, no regulatory or economic foreign policy tools were consistently deployed to foster the energy transition or support the US green industry abroad.

The arrival of Obama resulted in major changes in the US green foreign energy policy. Unlike in China and the EU, the US presidential political cycles, coupled with the continuous polarization of climate issues, produce abrupt fluctuations in policy, although these are sometimes constrained by the role of Congress. Under Obama, the US relaunched its multilateral climate diplomacy, contributing to the Paris Agreement. However, Obama also improved the US minilateral and bilateral diplomatic engagements in the area of climate change, renewables, and low-carbon technology. Moreover, he increased the US climate and development aid financing for promoting abroad not only regulatory and market reforms, in line with US preferences, but also renewable energy. As in the case of the EU, this green soft power—channelled, for example, through the programmes managed by the USAID, which also rely on capacity building and technical assistance—aimed to indirectly support the diffusion of renewable energy or investments in electricity networks by shaping third countries' regulatory and policy frameworks. Still, mirroring the post-2008 shift towards more interventionist domestic approaches, under Obama, the US intensified the use of economic tools for its green foreign energy policy and expanded its action in the area of industry, investment, and trade. This move was also driven by the changing international landscape, especially the growing role of Chinese investments and companies in several regions of the Global South. These economic tools were deployed by Washington in a catalytic power fashion, as in the case of the EU, but also according to a green economic power strategy. Indeed, although the US developmental state approach is different from that of a quasi-state like the EU, Washington also used network diplomacy and financial mechanisms to create multi-actor coalitions in order to better pursue its green foreign energy policy goals. A catalytic power approach also resonates with the US liberal domestic setting and its views on state–market relations. However, unlike the EU, the US could leverage its armamentarium for foreign economic policy and commercial diplomacy, such as through the OPIC and EXIM, to support (globally) the internationalization of US green business and reinforce relations with other countries through financial involvements. In turn, these latter facilitated the diffusion of renewables (and US green technologies) abroad in a more direct way, promoting the realization of wind and solar projects. Yet, this manifestation of the external reach of the US developmental state approach was very different from that of China, reflecting the diverse features of their national models. It was built around a major role for private capital rather than SOEs or state-owned policy and commercial banks. Moreover, the US, like the EU, used its economic tools as vehicles for promoting in third states those norms and principles that were aligned with its preferences.

The increased use of economic tools under Obama encountered several internal constraints, as exemplified by Congress blocking the EXIM's operations in 2015. Internal constraints also limited Obama's climate agenda at home and abroad, hindering US green soft power, which could not rely on leadership by example. A similar dynamic characterized Obama's efforts to relaunch the US international liberal agenda through transregional trade agreements, i.e., through the TPP and the TTIP. This strategy was a response to the growing difficulty in promoting global deals through the WTO, but the TPP in particular also had the goal of containing China's influence in the Indo-Pacific and promoting rules and standards different from those 'exported' by Beijing through the BRI. This strategy had a green regulatory power component, but it was abandoned with the arrival of Trump, which reversed Washington's international liberal posture and promoted important transformations in the US green foreign energy policy. Under Trump, the US disengaged from the multilateral climate regime and, overall, its green soft power diminished. This reflected changing preferences in the presidential agenda, in which climate change and the energy transition were marginalized. However, thanks to the constraints imposed by Congress in this period, some building blocks of the US green foreign energy policy survived. The growing rivalry with China even led to a relaunching of the US economic tools, as exemplified by the upgrading of the OPIC into the DFC and the EXIM's Program on China and Transformational Exports. This rivalry also spurred increased attention on critical minerals and REEs. These innovations, however, were not part of a comprehensive green power strategy, although Trump's trade war with Beijing extended to green technologies.

The US green foreign energy policy changed again with the arrival of Biden in the wake of the Covid-19 crisis and a further deterioration of the international landscape triggered by the war in Ukraine. In line with the new administration's preferences, the US placed climate action and the energy transition at the centre of Washington's foreign policy and global strategy. It relaunched all US diplomatic engagements on climate change, renewable energy, and low-carbon technology and created new institutional venues for channelling US green soft power. It increased US climate financing and prioritized green foreign energy policy objectives in Washington's development assistance. Even more markedly than under Obama, the shift in state–market relations towards more interventionist approaches at home was paralleled by efforts to upgrade the external reach of the US green developmental state.

Due to the growing rivalry with China, this upgrade acquired a more strategic dimension, especially with regard to the security of critical low-carbon supply chains. In this latter case, a major emphasis was placed on bilateralism and minilateralism with US allies and actual or potential partners. The Obama era's international liberal agenda was replaced by the 'new Washington consensus', as well as by a mercantilist approach based on a more muscular deployment of US economic power. This change in preferences is visible in the new US regional and bilateral agreements, which do not aim to increase trade liberalization or create binding international institutions but to coordinate green industrial policies in a more flexible way and promote investments that increase the resilience of low-carbon supply chains. To achieve this goal, the US has resorted to a direct mobilization of its economic might and commercial diplomacy, for example, through DFC investments in critical minerals and manufacturing projects abroad—and even to catalytic power and network diplomacy to create larger coalitions with public and private partners. The US mercantilist turn, on the other hand, is visible in several IRA provisions. Unlike the EU, Washington can adopt a more muscular green power strategy to increase its industrial capacity and reduce its dependency on China. Finally, under Biden, the US has linked its green foreign energy policy objectives with its security and military alliances and partnerships. Indeed, unlike China—and of course unlike the EU—Washington has the capability to develop a green power strategy that is sustained by its global military posture. This is particularly the case in the Indo-Pacific, where the challenges posed by the Chinese rise are more pressing.

Limitations and potential of Chinese, EU, and US green power strategies

The limitations and potentials of the Chinese, EU, and US green power strategies are related to their internal capabilities and weaknesses, as well as to the constraints and opportunities offered by an evolving international landscape. China's green power strategy, which is mostly based on economic power and close coordination of the state–industrial complex domestically and abroad, has been effective in supporting the country's rise to the top of the green technology ladder. Not only is China a technology maker in many relevant sectors of the global energy transition, but it has also extended the reach of its external influence thanks to its achievements in the area of renewables and low-carbon technologies. These achievements are sustained by its developmental state model, which will help China to maintain the lead in key green sectors. According to the IEA's (2024) projections, in the next five years, China will deploy almost four times more renewable energy capacity than the EU and five times more than the US (they will remain, respectively, the world's second and third largest growth markets). Moreover, considering the existing and planned projects, in 2030, Beijing is projected to still hold a prominent position in the manufacturing of solar PV (79% of global capacity), wind technology (64% of global capacity), batteries (68% of global capacity), and electrolysers (26% of global capacity; IEA 2023c). By that year, the EU and US manufacturing capacity is also expected to have increased as a result of recent efforts on the matter, but neither will challenge China's leading position. For example, by 2030, the EU and US share of global wind manufacturing capacity is projected to be, respectively, 15% and 8%, the share of global battery capacity, 11% and 15%, and the share of global electrolyser capacity, 24% and 21% (IEA 2023c). Similarly, China is expected to maintain a very strong position in the refining and processing of several critical minerals and REEs required for the energy transition. These industrial and technological capabilities continue to present a solid foundation for the Chinese green economic power strategy, alongside the international activities of its companies (both SOEs and private companies) and Beijing's financial capacity and commercial diplomatic outreach, including through BRI engagements.

However, this success has triggered counter-reactions that might gradually worsen the international environment for China's external influence. Both the EU and the US have designed measures to reduce their dependency on China, expand their industrial capacity, and promote low-carbon supply chains that are less centred on Beijing. In these processes, they have involved several other countries from both the Global North and the Global South. In the short term, these efforts will not drastically challenge China's position, but in the longer term, they might erode the base of China's green economic power and promote a de-concentration of low-carbon supply chains. Indeed, although the Chinese developmental state (and party-led) model, with its close state–business alliances, possesses important strengths, it also presents limitations in the context of growing competition among the great powers. The most obvious are related to the way in which Chinese companies and investments are seen in Western countries, notably in the US. These issues go beyond the specific IRA restrictions on FEOC. Despite growing geopolitical tensions, after the launch of the IRA, some Chinese companies planned to build new batteries and solar manufacturing facilities in the US. A very high-profile example is a $3.5 billion partnership between Ford and the Chinese battery giant CATL. Although Ford has stressed that 'zero tax dollars will go to CATL' through this partnership, the plant has been banished from the state of Virginia, and the deal has created mounting political opposition from both Democrats and Republicans (McCarthy and Olano 2023). Other large projects involving Chinese companies—e.g., solar manufactures such as Longi, JA Solar, and Hounen Solar—are proceeding, but political tensions between Washington and Beijing could similarly put these investments at risk. These tensions could also affect the EU's approach to Chinese investments and green technology, although in Europe the positions on how to tackle Beijing's rise are more diversified.

Another sensitive area is trade, in which Chinese prominence is increasingly perceived as a potential threat. In this case, too, so far, the EU has taken a more cautious approach than the US, with its tariffs targeting Chinese solar technology and EVs. However, the anti-subsidy investigation into the imports of battery electric vehicles from China launched in late 2023 by the European Commission might represent a change of direction in this regard. Discussion of additional tariffs on imports of EVs from China—currently a 27.5% tariff on EVs from China is in place in the US, whereas in the EU, the import duty is 10% (European Parliament 2023)—began in Washington on December 2023 (Duehren 2023). Tesla CEO Elon Musk even warned that without trade barriers, Chinese EV companies would outcompete most other automakers (Roy 2024). In both cases, however, the final results of these initiatives are still uncertain; 2024 is an electoral year in Washington and Brussels, and industrial interests in Europe and the US are divided on the matter. Moreover, similar decisions could risk slowing down progress towards their energy transition objectives by rising the costs of deploying low-carbon technologies. In addition, China has several forms of leverage for retaliation in the case of trade disputes, including through its control of many critical minerals and green technology.

China, however, is also diversifying its trade relations in the green realm, especially towards the Global South, including in new regions such as the Middle East. At the same time, an increasing rivalry with the West, coupled with the very nature of the Chinese regime, could also limit Beijing's influence in some countries. For example, Argentina reversed its position on joining the BRI after the recent election that resulted in the victory of the far-right-wing President Milei. Nonetheless, the Chinese approach, which focuses on promoting investments and projects instead of rules, has been very effective, especially in the Global South, at sustaining the expansion of the Chinese sphere of influence, which now also structurally includes a green dimension. As China's green industrial and technological capacity continue to increase—and as the global energy transition proceeds—this sphere of influence has the potential to further consolidate, not only via trade and investments but also through new industrial partnerships in the mineral, EV, hydrogen, and battery sectors. Conversely, Chinese green soft power faces significant obstacles. On the one hand, China's green developmental state model can be an example for those developing countries willing to expand their industrial footprint in the emerging low-carbon world. On the other hand, China has not embraced the role of climate leader, and its coal-based energy system still represents a constraint on action in this field. Moreover, China, due to its preferences and views on the matter, has not emerged as an exporter of policies or new ideas capable of shaping the energy transition; the EU's ETS or the USAID-led auctions for renewable energy installations stand in marked contrast. In addition, the peculiarities of the Chinese authoritarian model, centred on the CCP, can hinder its potential 'leadership by example', especially in a context in which (green) industrial policies have become the new normal, even in the Global North, as illustrated by the 'new Washington consensus'.

The Chinese green regulatory power, which as we have seen is a less developed component of Beijing's overall approach, faces similar problems. In this case as well, because of its preferences and domestic structures, China has not sought to expand its influence by creating regulatory regimes, nor has it used its trade power to promote rules and standards favouring the energy transition abroad. Even the diffusion of Chinese green technological standards has proceeded mainly through the 'hub and spoke' model enshrined in the BRI. This situation has started to change recently, but in order to become a true rule maker, Beijing has to improve its regulatory capacity in the sectors relevant to the global energy transition. Moreover, in the sensitive area of standardization, both the EU and US have been moving towards a more strategic approach that could undermine Chinese efforts to turn into a global rule maker.

The EU, over time, has developed a flexible toolkit for green foreign energy policy, but its actorness and effectiveness (e.g., Bretherton and Vogler 2005, 2013) are limited by strategic incoherencies, an internal capabilities gap, and a more competitive international landscape that is an obstacle to its traditional multilateral approach to energy, climate, and trade. As explained, the EU green power strategy is based on a combination of soft, regulatory, and catalytic power. However, despite recent efforts, the EU has not yet fully developed its green economic power, and several traditional tools of economic statecraft remain outside the competencies and scope of its actions. This gap in capabilities characterizes even its internal policies and has proven significant as the bloc shifts towards forms of 'domestically oriented' economic statecraft. As shown, the EU green developmental approach has taken the form of a catalytic state. The emergence of this institutional structure has favoured more focused and coordinated industrial policies across Europe, which has delivered some important results (e.g., Terzi, Sherwood, and Singh 2023). However, it is not clear whether this model—at least without significant changes—will be able to sustain increased industrial and technological competition with China and the US. For example, in terms of amount of subsidies for EVs, green manufacturing, and renewable energy, the EU actually offers more than the US does through the IRA (Kleimann et al. 2023). However, the system of EU incentives and measures is much more fragmented and complex than the simple and effective IRA tax break provisions. In addition, EU subsidies do not discriminate against foreign producers in the way the IRA does (ibid.). These limits can pose a challenge to the ambitious targets set with the NZIA and the CRMA. Moreover, by favouring the use of state aid by member states over building stronger EU-level financial capabilities—a path that has so far been opposed by several national governments—this approach risks exacerbating imbalances within the single market. In the short term, member states can nurture their own national industries, but the overall competitive stance of Europe risks being undermined, as even larger member states do not have the resources to equal the other major powers. Another problem in the EU's efforts to reduce dependency on China in key low-carbon supply chains is related to the generally low level of public acceptance for mining activities across Europe (e.g., Tennant 2023). This may cause delays in reaching EU targets in the area of critical minerals and add complications to projects that already require years to be fully operative.

In general, divisions among the member states also limit the definition of a coherent approach towards China. First, there are significant differences between Western and Eastern EU member states, and many of the latter (i.e., Croatia, Czech Republic, Estonia, Hungary, Latvia, Lithuania, Poland, Romania, Slovak Republic, Slovenia) have joined the BRI, showing less concern about this vehicle of Beijing's influence on the continent. Divergences also exist among Western European countries. For example, Italy joined the BRI in 2019, although it eventually withdrew in 2023. Moreover, there are various views on how to tackle China's green rise; France often pushes for strengthening the EU's industrial defences against Beijing, while Germany, which is heavily exposed to the Chinese market, sustains a more cautious approach. All these divergences can undermine the EU's capability to act as a coherent actor, especially when more muscular and/or protectionist measures in line with a green economic power perspective are on the table. They also cast many doubts on the actual ability of the EU to uphold its geopolitical turn. Finally, differences between the EU approach and those of the member states can also hinder the bloc's green foreign energy policy. For example, after the Russian invasion of Ukraine, member states' international activism in the area of natural gas and LNG increased widely, but unlike in the case of the EU, only a limited portion of their new bilateral deals in this field included a green element. In particular, from February 2022 to August 2023, the member states and their companies signed thirty new gas deals with third states, but only seven contained references to renewable energy (ECFR 2023). Moreover, whereas the EU hydrogen strategy places a strong focus on green hydrogen, several member states, which have been developing their own national hydrogen strategies, began cooperation with fossil fuel exporters in the area of blue hydrogen (i.e., hydrogen produced from natural gas using carbon capture and storage technology). This cooperation especially targeted MENA countries like Algeria (e.g., deals with Italy) and Gulf monarchies, such as the UAE and Saudi Arabia (e.g., deals with France, Germany, and Italy). Conversely, EU external engagement on the matter with the Gulf Cooperation Council's countries has remained limited, partially due to this strategic incoherence (Bianco 2023).

At the same time, and despite these constraints, the EU has been able to upgrade its green catalytic power, which can, with its focus on projects, investments, and industrial partnerships rather than rules, potentially augment the bloc's influence in several developing countries. It can also help to sustain the shift towards minilateralism and bilateralism in EU foreign engagement owing to the deterioration of the global multilateral landscape. Moreover, this approach can be deployed more easily in coordination with other 'like-minded' Western countries and the US, helping the EU to fill its capabilities gap. Green catalytic power can also deliver geopolitical results if it allows the EU to mobilize sufficient resources to counter China's economic power in the Global South or helps the bloc diversify low-carbon supply chains. For example, the EU, through its Global Gateway, is joining forces with the US and the PGII to promote the Trans-African Corridor. This project should enhance connections between the DRC and the so-called 'Copper Belt' in Zambia to Lobito port in Angola, foster investments for the processing of the extracted raw materials, and increase access to world markets for the products from these regions (European Commission 2023c).

However, the resources allocated to the EU financial tools for external action are still narrow, especially beyond the EU neighbourhood and Africa. Moreover, the EU blending and guarantees operations are embedded in the rules that guide the bloc's official development assistance, which are aligned with the OECD framework. This means that the investment projects supported by the EU and member states' development banks should both adhere to high environmental and social standards and consider the financial sustainability of the aid. In addition, the EU (largely) follows the OECD recommendation on 'untied aid' (e.g., OECD-DAC 2022); consequently, its financial assistance can be used to support the internationalization of European companies only to a very limited extent (European Commission 2023d). That is to say, the mandate of the EU development banks, as well as of the development banks of member states—unlike that of their export credit agencies—focuses primarily on developmental goals rather than on promoting national business. On the one hand, all these constraints reduce flexibility in the use of these tools in pursuit of the EU's foreign policy interests. On the other hand, they help the EU promote abroad, together with investments and projects, specific norms and principles for shaping the global energy transition. Moreover, in a less direct way, the EU's financial tools can contribute to the internationalization of European green business; this action, however, resembles network diplomatic practices more than traditional commercial diplomacy. For example, several wind and solar projects supported by EU blending facilities in the EU's neighbourhood and Africa were realized with equipment from European manufacturers, such as the Wind Farm Gulf of Suez in Egypt (wind turbines from the German Siemens) and the Lake Turkana Wind Farm in Kenya (wind turbines from the Danish Vestas). This is also the case for the projects under the CEF-E: the ELMED and the EuroAsia Interconnector, for example, involve European electricity companies such as ENEL (Italy) and construction companies such as Nexans (France).

In this regard, it is also worth noting that European companies have been important drivers of renewable energy projects worldwide. In Latin and Central America, for instance, 75% of the last decade's foreign direct investment in renewables projects came from European companies (mainly Spanish, Italian, and French power companies; Escribano and Arbeloa 2023). The Italian electricity utility ENEL Green Power has developed approximately 11,000 MW of wind and PV projects in this region (ENEL 2023), more than double that developed by Chinese companies; and the Spanish Iberdrola, through its local subsidiaries, has installed around 3,000 MW of wind and solar power in Brazil and Mexico. European foreign direct investment has been characterized by high labour, social, and environmental standards and a high level of technology transfer (e.g., wind turbine manufacturing plants in Mexico and Brazil), in contrast to the activity of investors from China and the US (Escribano and Arbeloa 2023). European power companies are also important players in Africa; for instance, ENEL Green Power and the French company Engie have a portfolio of around 3,000 MW of solar and wind installations on the continent (ENEL 2023; Engie 2023). If the EU were better able to exploit the complementarity between the technological and financial capacity of European companies and its green foreign engagements, the overall influence of the bloc would be amplified. As mentioned, this idea has begun to take root with the proposal for an EU-level export credit agency, but effective steps in this direction are still lacking.

The area of regulatory power is more traditional for the EU, which can count on important capacity and competences in this field. Regulatory power is also aligned with the long-standing identity and preferences of the bloc, and the EU has been able to integrate several countries in its close neighbourhood into common regulatory regimes, including states with important potentials for the energy transition such as Norway. By accompanying electricity grid connections with regulatory harmonization and market integration, the EU has also been able to extend its regional influence. However, this approach has several limits when the preferences of the other states are not already in line with those of the bloc, as illustrated by the problems encountered by the EU in extending its regulatory outreach in the Southern Mediterranean. The EU can also leverage its trade deals to pursue green foreign energy policy objectives; however, there are again important limits. For example, the EU was able to reach a deal with Chile in 2022 that covers green hydrogen and critical raw materials (such as lithium and copper), yet the bloc has not been able to finalize a similar agreement with Indonesia, the world's largest nickel producer—precisely because of its possession of this valuable resource for the energy transition, Indonesia is acquiring growing economic leverage. In 2020, the EU also launched a dispute under the WTO against Indonesia for its export restriction and domestic processing requirement on nickel. Although in 2022 the WTO ruled against Indonesia, it is not yet clear whether the EU will be able to influence Jakarta's policy on the matter. Moreover, the EU has no free trade agreements in place with other important players in the area of critical minerals, such as Australia, the DRC, or the Philippines. Also, the EU–Mercosur Comprehensive Trade Agreement is still awaiting ratification. The political dynamics behind this agreement reveal other potential problems of the EU approach. On the one hand, as Mercosur countries hold vast reserves of the critical raw materials crucial for the EU energy transition, it will help the bloc reduce dependency on China. On the other hand, this agreement has been criticized by several environmental NGOs, members of the European Parliament, and national governments for its negative effect on climate change and sustainability. In addition, it has also been criticized by agricultural organizations across Europe for its foreseeable negative impact on the sector, which has further reduced the willingness of large member states, such as France, to finalize it. Indeed, domestic politics is another rising constraint to the EU international trade agenda, as the appetite for protectionist measures has increased in recent years. As trade agreements have long represented an important vehicle for the EU's external influence, and as other tools, such as those of green economic power, are still underdeveloped, the final result of this trend could be a diminished role for the EU in the global energy transition. In this vein, measures like the CBAM face fewer domestic constraints, but it has spurred several criticisms for its protectionist potential among the bloc's trade partners (including from China). This despite the EU sought to design this measure in accordance with WTO rules. The CBAM also sets ambitious goals related to promoting energy transition measures and decarbonization abroad, favouring alignment with EU rules in several countries, and generally expanding the global reach of the EU green regulatory power. Work towards these goals has already begun, and various EU trading partners have enhanced their climate policies (or plan to do so) in response to the CBAM (Heung, Rydge, and Chan 2023). The actual effectiveness of this initiative in the near future, however, will also be influenced by the level of support it receives from other major international players (e.g., Sabyrbekov and Overland 2024). The green regulatory power wielded by the EU has also shown potential in other sectors, although with less visibility. In the area of green hydrogen, where the EU would emerge as an importer, other countries are drawing on the EU's sustainability requirements to design their own regulatory framework (Zabanova 2023). In addition, unlike China, the EU has important allies in the US and other industrialized countries who are willing to work with the bloc to set global standards for emerging sectors of the energy transition. For example, since 2022, the European Commission and the US DoE have promoted cooperation between the European Battery Alliance and the US Li-Bridge Alliance to jointly develop solutions and standards for the supply chains of lithium-ion and next generation batteries, including in the critical materials segments.

The abovementioned initiatives—additional examples might include the EU green taxonomy, the Climate Club, and the Net-Zero Industrial Partnerships—suggest that the EU continues to have the potential to deploy green soft power by fostering new ideas and policies and by leading by example. These resources can help the bloc exercise influence in the emerging low-carbon world, despite its declining economic weight and reduced CO_2 footprint. Although this traditional channel of EU influence is increasingly being constrained by the rise of China (and other emerging powers), the most challenging obstacle for the EU green soft power approach, in the short term, is domestic politics: multiple energy transition and climate measures, including those approved under the EGD, face growing opposition within Europe. Forecasts for the 2024 European Parliament elections indicate a major shift towards the right in many EU member states, with populist far-right parties gaining votes and seats and centre-left and green parties losing votes and seats (Cunningham et al. 2024). If confirmed, such political change could result in a reversal or downscaling of the ambitious EU climate and decarbonization agenda, which in turn might hinder the bloc's green soft power.

Domestic (political) constraints present similar limitations for the US green power strategy. With the enduring polarization on climate change—which remains a low-priority issue for voters who identify as Republicans or lean towards the Republican Party (Tyson, Funk, and Kennedy 2023)—Washington's green soft power is contingent on the results of the presidential and/or congressional elections. These can provoke abrupt shifts in US engagement in the multilateral climate regime and in its bilateral and minilateral energy diplomacy. Moreover, even when (comparatively) pro-green presidents are elected, their more ambitious actions and plans can be halted by Congress. For example, at the COP 28 in Dubai, the Biden administration made a $3 billion pledge to the Green Climate Fund, but it is not clear whether Congress will support such a decision. Overall, domestic politics limit the US potential to lead by example and create persistent uncertainty about the credibility of Washington's commitments on climate mitigation and decarbonization. At the same time, some components of US green soft power have shown a certain resilience, even under a climate sceptical presidency like that of Trump. The US foreign policy and foreign aid apparatus (e.g., USAID) maintains important resources to promote and shape the energy transition abroad, especially when sustaining renewable energy can be coupled with market-based policy ideas and reforms. These resources can be amplified under pro-green presidents, and in these cases, the US has demonstrated significant capabilities and ingenuity in creating new institutions and channels to increase its role in the global energy transition. In the Global South, however, US green soft power is increasingly being challenged by China. Although Washington has reacted by more enthusiastically presenting an alternative model to the Chinese state-led green approach, it has become easier for countries in the Global South to turn to Beijing for assistance and technological solutions.

Domestic constraints have also had negative effects on US green regulatory power. As illustrated, with the NAFTA and then the USMCA, the US has built a regional regime with relevance for renewable energy, electricity, and low-carbon technology. However, since Trump's withdrawal from the TPP, trade liberalization has been sidelined in the US agenda. The 'new Washington consensus', with its focus on industrial policy and non-binding modes of international cooperation, can be attractive for countries in the Global South. Yet regional trade deals also offer advantages; they can increase commercial exchanges and technical and regulatory cooperation in the area of renewables and low-carbon technology and can extend the influence of larger economies. In the Indo-Pacific, for example, by backing the RCEP, China can further consolidate its role and linkages with regional actors, including US partners and allies. Washington's retreat from its traditional international liberal strategy might, in the long run, undermine its global posture, even in the green realm.

The rivalry with China, however, has been a driving force in the development of the US green catalytic and economic power strategies. This was already visible under Obama and escalated with Biden after an uncertain phase during the Trump presidency. In the area of catalytic power, the US, like the EU, has important potentials, as it can leverage its diplomatic outreach and economic tools to build multi-actor coalitions that involve other countries as well as private partners. To a greater extent than the EU, the US can rely on its global status and significant financial resources to pursue its foreign policy objectives in this way. Moreover, unlike China, which deploys its economic might through bilateral modes of engagement and relying on its own capacities, the US can use these (minilateral) coalitions to co-opt its major allies in different regional theatres, many of which (e.g., the G7 members) have the financial resources, expertise, and willingness to promote the global energy transition and to counter Beijing while promoting investments and infrastructural projects abroad. On the other hand, catalytic power approaches, like those foreseen by the PGII, which count on the possible mobilization of large amounts of private capital, risk missing their expectations. From the perspective of countries in the Global South, and in many cases of their ruling elite, the promise of such resources can be seen as less tangible and desirable than, for example, the direct investments of Chinese companies or loans from Chinese banks. This can limit the overall international influence that can be obtained through these vehicles. Moreover, like the EU, the US catalytic power approaches maintain a focus on promoting certain rules and principles, which can limit their flexibility and encounter opposition in third states.

In the area of green economic power, the US has recently realigned its armamentarium to play a more effective role in the global energy transition and better pursue its national interests, both domestically and abroad. Washington's green economic power can be effective, especially if strategically combined with its military reach. This might help the US to create and reinforce strategic (green) alliances and design spheres of geopolitical influence in the emerging low-carbon world. This potential, which is similar to that possessed by the US in the fossil fuel–based global economy, remains a crucial difference between the US and the Chinese and European approaches. However, perennial domestic political constraints and a declining role in some regional theatres might hinder this component of the US green power strategy.

Under Biden, the US has, unlike the EU, embraced a more protectionist and direct approach to expanding its green industrial capacity and restructuring low-carbon supply chains according to its interests. With the IRA, the US developmental state approach has escalated traditional industrial policy measures—e.g., targeted subsidies and tax credits to increase manufacturing, investment, and demand—that are delivering important results. As of January 2024, for example, $120 billion in investments in new battery manufacturing and supply chains in the US have been announced, including over two hundred new or expanded minerals and materials processing and manufacturing facilities. Another $35 billion has been announced for new EV component and assembly plants, $16 billion for solar manufacturing facilities, and, respectively, $3.5 billion and $1.6 billion for offshore wind and electrolysers and fuel cell manufacturing facilities (US Department of Energy 2024). The IRA has also successfully attracted several investments from foreign firms—US-based companies accounted for less than half of the green energy manufacturing announcements made in the first year after the passage of the IRA and only approximately one-third of the promised investment—including from South Korea, Japan, Canada, and European green leaders such as Germany (McCarthy and Olano 2023). The IRA will also deliver important climate results. Estimates suggest that it could help the US reduce emissions from 43–48% below 2005 levels by 2035, in comparison to the 27–35% reductions foreseen in non-IRA scenarios (Bistline et al. 2023). Moreover, although the Chinese global prominence in several segments of key low-carbon supply chains will endure in coming years, IRA provisions such as those on FEOC can foster a restructuring of these supply chains, improve diversification, and support the US's 'de-risking' strategy. To accomplish this goal, the US has also begun to mobilize its foreign economic policy apparatus, as highlighted, for example, by the DFC investments in green manufacturing and critical minerals projects abroad. In addition, the IRA has triggered a new round of green industrial measures, not only in the EU but also in several other countries, including Japan, India, the UK, Canada, and South Korea. If sustained, these efforts, which are driven by close US allies or democratic partners, might result in further diversification of low-carbon supply chains, a process that, in turn, can result in the long run in a global industrial energy transition landscape that is less centred on China and more in line with the interests of the West (assuming Western countries coordinate their policy and national measures). Finally, Washington has the potential to leverage its global network of alliances and security partnerships to uphold its green economic power strategy. In the Indo-Pacific, these connections are especially evident, as the rise of China has heightened worries among other regional actors. Thanks to its military posture, the US can cement its green relations with other countries and maintain spheres of geopolitical influence in the emerging low-carbon world. At the same time, this also implies that the declining role of the US in some regional theatres, such as the Middle East, can hinder its green economic power strategy.

Like the other components of the US green power strategy, its economic power must also face domestic constraints, although the rivalry with China makes its core more difficult to dismantle. Yet the reversal of climate-related measures can hinder US green efforts, even in the areas of industrial policy or supply chain diversification. Moreover, less willingness to deploy economic tools in US international engagements—in line with traditional anti 'big government' attitudes—can hamper the achievements of important green foreign energy policy objectives. For example, in 2024 the Biden administration requested Congress increase the budget for the DFC and expand its ability to make equity investments (with an allocation of $2 billion) as part of its Out-Compete China Initiative (The White House 2024). This move would facilitate US efforts to improve the diversification of low-carbon supply chains, support investments in infrastructures, and counter Chinese BRI and overseas financial engagements. However, it is not clear whether similar proposals will gain political support in Congress. Further, there are two trends that might increase the resilience of the US energy transition and create positive spillover into Washington's green foreign energy policy and green power strategy. The first is linked to the impact of the IRA on the national industrial landscape. Indeed, although Republicans did not back this piece of legislation, it has spurred investments in many US states, including 'red states', i.e., states in which a majority of voters elect members of the Republican Party. These investments in manufacturing and other green-related businesses can augment the pro-green political coalition and increase the costs of dismantling the measures in this field. The second important trend, accelerated under Biden, is the growing interest of the military and the DoD in renewables and low-carbon technologies. This development can also reinforce the US energy transition and make related policies more resilient vis-à-vis abrupt changes in political majorities. In addition, as military equipment and technological innovations incorporate the shift towards renewables, the US may emerge as a global provider of such assets to its allies and potential security partners, further increasing its influence in the low-carbon world (e.g., Jaffe 2023).

7
Conclusions

State–market relations and green power strategies

We live in a world dominated by fossil fuels, which are still a key ingredient of the global political economy and international security. The shift towards renewables and low-carbon technologies, however, has accelerated in recent decades, driven by a combination of factors that include climate mitigation efforts, energy security considerations, and economic and geopolitical imperatives. A new green world order is slowly but steadily emerging—international affairs are becoming more and more embedded in a global energy system in which renewables have abandoned their niches to diffuse across sectors and markets. States are increasingly factoring these sources, the technologies needed to exploit, store, and transport them as well as their supply chains, into their foreign policy calculations. This trend is unfolding alongside the erosion of the Western-dominated liberal order and the return of geoeconomic and geopolitical rivalries among (capitalist) great powers.

Against this background, I wanted with this book to contribute to filling a gap in the burgeoning literature on the geopolitics of renewables and low-carbon technology and/or of the energy transition. Unlike several prior works on the matter, which focused on macro dynamics and scenarios or micro-level empirical studies, I explored the conceptual middle-level represented by the green foreign energy policies of major international actors. I developed an analytical framework to map this composite policy field and allow scholars to trace and explain changes over time as well the similarities and differences among cases. I applied this framework to the current three green superpowers: China, the EU, and the US. These three players possess the necessary assets to exercise a growing influence in the global energy transition and to expand their international posture through their green foreign energy policies and green power strategies.

In this conclusive chapter, I want first to return to two of the key themes of the book: evolving state–market relations in the area of renewables and low-carbon technologies and the internal–external nexus in green foreign energy policies and green power strategies. In the following section, I highlight the

Green Superpowers. Andrea Prontera, Oxford University Press. © Andrea Prontera (2024).
DOI: 10.1093/9780191987304.003.0007

opportunities and risks that these strategies imply. Finally, I discuss possible future scenarios for the emerging green world order.

As highlighted throughout this book, state–market relations are crucial to understanding the structural features of each green superpower's approach to the global energy transition. State–market relations are a constitutive component of state capacity, and these capabilities enable or constrain the green superpower's actions at home and abroad. State–market relations are embedded in national institutional and ideational configurations, but they are not fixed—they change over time in response to both external and internal challenges. All three green superpowers have expanded the scope and depth of their intervention in the green realm since the 2008 financial crisis. Whereas for China this constituted an upgrading of its industrial and technological growth trajectory, for both the EU and the US, the transformation was more significant. As these internal shifts reverberated into their external action, their green power strategies were affected as well. China's green power strategy is understandable only when considering the close state–market relations of its state–industrial complex; overall, this internal–external nexus offers a fundamental entry point for analysing and explaining patterns of similarities and differences among countries.

In recent years, state–market relations have continued to evolve in response to the even more challenging international environment caused by the Covid-19 pandemic and the tensions created by the Russian aggression towards Ukraine. Europe and the US have seen domestic developments as well, including the emergence of different political majorities, and these have contributed to changing the balance between states and markets. However, a general trend towards growing government interventionism is clear. This is particularly evident in the area of green industrial and technology policy, where forms of 'domestically oriented' statecraft have impressed a strategic turn on each green superpower's developmental state approach. Carbon neutrality pledges have also contributed to growing state interventionism, although in this case mounting political resistance to ambitious climate action functions as an effective constraint, and policy reversals are more feasible. At the same time, fundamental differences among the Chinese, EU, and US forms of economic, industrial, and energy governance have resulted in distinctive state–market relations, which (again) have been reflected in their green power strategies. The Chinese state–industrial complex, which grants a key role to state capital, continues to influence the various forms and modes of Beijing's green (economic) power approach. The same applies to the EU, the quasi-state nature of which affects the patterns of state–market relations at home, and these in turn are mirrored in its peculiar green power strategy, in which catalytic and regulatory power approaches have undergone robust development in recent years, unlike economic power that has to face several constraints. State–market relations in the US are conducive to the use of green economic power, catalytic power, and regulatory power; the first has been widely expanded under Biden. However, US green economic power continues to reflect its domestic settings, in which private capital and companies play a prominent role. In other words, the peculiarities of state–market relations help to explain the structural contours of each major power's behaviour and approach towards the global energy transition as well as their potential strengths and weaknesses in this transformative process.

Overall, however, all these developments show how the green superpowers are increasingly mobilizing the capabilities at their disposal, including in terms of state–market relations, to augment their influence in the emerging low-carbon world and pursue the main goals of green foreign energy policy. Their competition increasingly transcends the internal–external divide and extends to the transnational space of low-carbon supply chains, which they attempt to structure or restructure according to their geopolitical preferences. This process is very difficult because these supply chains emerged and were consolidated during a phase in which renewables and low-carbon technologies were not a matter over which great powers needed to compete, as the liberal international order sustained by the US hegemony facilitated global exchanges and industrial complementarities based on a primarily economic logic. Reversing such logic implies a further reconfiguration of state–market relations in the global energy transition. This process extends well beyond the three green superpowers, as the shift from fossil fuels towards renewables and low-carbon technologies has made room for other countries to design national trajectories of green industrial growth. In fact, the green power strategies of China, the EU, and the US all incentivize such development. As they seek to structure and restructure global low-carbon supply chains—from green manufacturing to critical minerals and new energy carriers like hydrogen—they are increasingly favouring the spread of green developmental state approaches worldwide. In addition, unlike the post–Cold War neoliberal decades, this ongoing global push to redefine state–market relations is occurring in a permissive environment. The US, the former hegemonic power, has moved away from its liberal international posture to embrace more interventionist measures, as captured by the 'new Washington consensus' parlance. A similar shift has occurred in other Western countries and US allies, including Europe, while China and the other emerging powers have traditionally adopted capitalist models that feature a high level of state interventionism. In sum, the global energy transition is unfolding amidst a redefinition of state–market relations, which the transition itself is bolstering. This has triggered a self-reinforcing process with wide transformative potential as well as new opportunities and risks.

Opportunities and risks in the emerging green world order

A green world order is a system of international relations and interactions among great powers in which renewables and low-carbon technologies play an increasingly important role as objectives targeted by foreign policies and as a means to achieve wider geopolitical and geoeconomic goals. This green order is emerging amidst a double structural shift; one is in state–market relations, and the other is due to the (relative) decline of the US hegemony and the accompanying return of great powers rivalry. This new global landscape presents both opportunities and risks. On one hand, the growing competition among great powers—and especially among the three green superpowers—to win the green technology race can boost investments and innovations in several sectors, with positive long-term effects on decarbonization and climate change mitigation. These competitive dynamics, coupled with climate and energy security motivations, can create the momentum essential to overcoming the potent inertia of the fossil fuel regime. The climate crisis is manifesting more vividly around the globe in extreme weather events that are becoming a new normal in many regions, including in the Global North. This makes it more and more difficult for governments around the world to delay the necessary actions or ignore the issue. Energy security concerns linked to oil and natural gas dependency are also on the rise as the Russia–Ukraine war continues and the Middle East experiences mounting tensions in the wake of the Israel–Hamas war, including Yemen's Iran-aligned Houthis' attacks in the Red Sea. Moreover, as mentioned, the competitive dynamics around renewables and low-carbon technologies can foster green industrialization in developing countries. These countries can potentially also use the competitive dynamics among great powers as leverage to obtain investments and technology, as well as to sustain their national green growth strategies beyond asymmetric relations of mere 'extractivism'. As documented in the previous chapters, each of the green superpowers have recently increased efforts to foster industrial partnerships in their green foreign engagements. This, in the long run, can produce a positive de-concentration of several low-carbon supply chains, which will make them more resilient to unforeseen events and crises as well as to politically motivated disruptions. This outcome, in turn, can reduce some of the risks related to the global energy transition. In addition, the growth and spread of green manufacturing and industrial capacity, both in the Global North and in the Global South, can create larger domestic coalitions in support of the energy transition and related climate mitigation measures.

On the other hand, the growing competition among green superpowers, and especially the rivalry between China and the US, might result in additional protectionist measures, disproportionate trade tariffs, or even open green trade wars. This would cause abrupt disruptions in the existing low-carbon supply chains and limit technological transfer and industrial partnerships, with additional negative effects on technological learning, costs, and innovation. In particular, pursuing new forms of green energy independence or technological sovereignty could hamper the global development and diffusion of renewables at the scale and pace that is needed to speed up the energy transition and avoid a climate crisis. A weaponization of dependencies along low-carbon technology supply chains might further put at risk decarbonization objectives around the world, as countermeasures to tackle this weaponization could lead to duplicate investments and a waste of resources that could be, for example, channelled to support the energy transition in the Global South.

In sum, as competition among the green superpowers is becoming the new normal in twenty-first-century world politics, it is essential to find a balance between these competitive tensions and the level of cooperation and economic interdependence that is required to accelerate the global energy transition and meet ambitious climate mitigation targets. Finding this balance, however, will be very difficult, as many uncertainties remain regarding the long-term political and economic effects of the recent crises, from the Covid-19 pandemic to the Ukraine–Russia war, in both industrialized and developing countries. The instability in the Middle East caused by the Hamas terrorist attack on Israel on 7 October 2023, the subsequent Israel–Hamas war, and the enduring tensions in the Indo-Pacific, where China's assertiveness over the Taiwan issue is rising, add additional complexity to this picture. In this vein, climate change efforts can contribute to finding such a balance. China has cautiously adopted a more constructive role in the multilateral climate regime and global climate policy since the Paris Agreement. It has also shown a willingness to maintain such a position, regardless of the commitments of the US. At the recent COP 28 in Dubai, both industrialized and developing countries finally openly recognized the need to transition away from fossil fuels in energy systems. Renewables are expected to grow significantly in the next decade, not only in China, Europe, and the US, which will maintain their lead in this field, but also in emerging economies, such as India, Brazil, and other countries of the Global South. This could possibly ease the next rounds of multilateral climate negotiations and create more incentives to achieve the necessary momentum required to meet the Paris Agreement's objectives.

At the same time, there are concrete risks that such momentum will not materialize. Indeed, many uncertainties remain regarding the real commitments of the great powers towards decarbonization. In the EU and the US, this process will be increasingly affected by domestic politics, as the structural transformations triggered by energy transition will increase resistance to serious climate action. In the US, politicization of climate change is another important challenge that can undermine sustained and continuous efforts at home and abroad. In Europe as well, the political consensus in favour of the energy transition and ambitious climate targets is diminishing. Far-right populist parties, in particular, have learned how to exploit the challenges inherent to the energy transition and the fears of public opinion to block or reverse more incisive measures. Moreover, despite the emergence of the narrative of a 'just' energy transition and the promises made in multilateral negotiations since the Copenhagen summit, industrialized countries are not providing the amount of financial resources and assistance that is necessary to help those in the Global South to sustain their mitigation and climate efforts. For China, it will be similarly difficult to overcome fossil fuel lock-in, particularly if energy security or industrial development considerations retard the reduction of coal usage. Moreover, Beijing has only recently begun to include international climate finance as a component of its green power strategy, which has so far focused on its national economic and foreign policy interests. Cooperation among Western-led multilateral development banks, such as the EIB, the EBRD, and the World Bank, and the AIIB and NDB on joint climate action has also been established. However, notwithstanding its growing presence in this area, China continues to view climate change as a Global North versus Global South issue. The final results of these dynamics could reframe renewables and low-carbon technology only as an arena for geopolitical and geoeconomic competition, a process that would reduce the space for cooperation among great powers, fragment low-carbon supply chains, and, in the end, slow down the global energy transition and leave the world vulnerable to climate crisis. This risk should not be underestimated, as the tension between Beijing and Washington is intensifying and appetites for protectionism and climate action reversals are on the rise in Europe and the US, alongside varied forms of national chauvinism. In Europe, energy security concerns will continue to be an important driver for promoting renewables and reducing dependency on imported oil and natural gas, especially if the Russia–Ukraine war endures. However, with its declining economic weight and climate footprint, the EU cannot alone be a decisive force for the global energy transition. This is especially true if domestic constraints limit its soft and regulatory power, which so far have been major components of its green power strategy.

What's next?

As repeatedly stated, this book was not focused on charting possible scenarios for the emerging green world order. However, drawing on the lessons learned in the previous chapters and the selected issues reviewed above, we can glean some insights on the matter. In 2024, there are important elections in the US and the EU. As in the recent past, the US presidential election will probably have an important impact on US green domestic and international policy. If Biden is reconfirmed—or another Democratic assumes the presidency—the innovations introduced under his administration might be consolidated and green foreign energy policy goals could be institutionalized in the Washington global strategy. Overall, the current US green power strategy would probably be further strengthened, as renewables, low-carbon technology, and climate change will all remain priorities in the domestic agenda. This would facilitate additional advances in the multilateral climate regime and possibly stabilize the results achieved in the last COPs. If a Republican president is elected, including Trump, the US energy transition and green power strategy will probably undergo significant changes. However, as happened during the previous Trump administration, this does not mean a complete dismantling of the US green trajectory. While climate action would be negatively affected, the renewable energy complex will guarantee a certain resilience to the US green developmental state and to some building blocks of the US green foreign energy policy. As green manufacturing is a growing industry in several US states and the military has embraced the shift towards renewables and low-carbon technology, abrupt changes would be difficult. However, the probable disengagement from the multilateral climate regime would destabilize the recent global efforts and make it more difficult to ease tensions in Global North–Global South relations. In that case, the US green power strategy will probably maintain its recently adopted more mercantilist stance, but without climate cooperation efforts to soothe frictions with other industrialized countries and China. At the same time, under either a Democratic or a Republican administration, rivalry with China will remain a driving force of the competitive dynamics related to the global energy transition. This could function as an additional factor in stabilizing and even expanding certain muscular components of the US green power strategy, including through closer connections with its military posture. Although the tone of this competition could be different under different presidents, US willingness to cooperate with its allies and partners to counter China's rise can be assumed, especially in the Indo-Pacific. With regard to the US military posture, it is worth noting that, as illustrated by its response to the 2024 Houthi attacks in the Red Sea, Washington maintains a leading role in ensuring the security of critical sea lanes that underpin the world's economy and trade. This component of the US's (declining) hegemonic power is still unmatched. Military tools are less important for achieving foreign energy policy goals in a low-carbon world than in one powered by fossil fuels. However, they are still useful for sustaining a wider green economic power strategy and reinforcing green alliances with allies and security partners, and this could grant an advantage to Washington in the coming years.

So far, EU green action at home and abroad has developed in a path-dependent manner, with a gradual increase in ambitions, resources, and tools. However, the 2024 election of the European Parliament could result for the first time in a reversal of this trend, especially if far-right-wing parties obtain many seats. Moreover, regardless of the outcome of this election, opposition to climate and green policies, including those launched with the EGD, is growing across Europe. This may provoke resistance from member states, not only in Eastern Europe but also in Western Europe, which could hinder the energy transition in the continent as well as the EU international posture on the matter. The EU's traditional 'leadership by example' in the climate and renewable energy realm could be particularly affected by such a development, but it could also have negative repercussions for EU green regulatory and catalytic power. Competition with China, and even with Washington, will probably remain an important driver of the EU green industrial policy and green power strategy, as well as of its efforts to achieve more autonomy over strategic low-carbon technology supply chains. However, as noted throughout this book, the EU has less capability than the other green superpowers to embrace more muscular competition and risks being caught between a rock and a hard place, especially if Washington decides to 'go it alone' with respect to its green confrontations with Beijing. Without more common or coordinated efforts by the member states, these weaknesses will be difficult to overcome. This is, however, another puzzle that is difficult for the EU to solve. As explained above, the current context of the global energy transition reinforces shifts in state–market relations towards more interventionist approaches. However, a 'return of the state' in a quasi-state political system like that of the EU could trigger divergent dynamics, including both calls to reinforce EU-level capabilities and (national) calls to reinforce member state–level capabilities, and this phenomenon could be further amplified by the ascendency of far-right wing parties throughout the continent. If a balance between these tensions is not found—if, for example larger member states are convinced that it is better to 'go alone'—the overall result could be (at best) a business-as-usual compromise with limited impact on the overall ability of the bloc to act coherently and effectively in the area of green foreign energy policy. On the other hand, because of its limits, the EU will have important incentives to cooperate with the US and China to tap capability gaps and achieve wider foreign policy objectives. This could create problems in transatlantic relations if the China–US rivalry further escalates. At the same time, the EU could also emerge as a 'fixer' that alleviates the competitive tensions of the global energy transition. This is because its weaknesses can turn into opportunities to maintain multilateralism and a higher level of integration along renewable and low-carbon technology supply chains. However, this role as a 'fixer' is also likely to face domestic constraints, as multilateralism and trade openness are being challenged within Europe itself, and the appetite for protectionist measures is on the rise.

Unlike in the US and the EU, China's green trajectory will probably continue with fewer uncertainties. Renewables and low-carbon technology have long been part of the country's development strategy and, recently, have been integrated into its overall foreign relations. China has much to gain from the global energy transition, both in terms of energy security and of economic and commercial growth. Moreover, Beijing can continue to leverage its green industrial and technological capabilities to increase its influence in the Global South. The challenges for China will probably derive more from the changing international landscape, especially regarding the measures enacted by the US and EU to reduce China's centrality in several low-carbon supply chains and those to limit its exports. These initiatives, however, are unlikely to result in a decline in China's global influence in the emerging green world order, as Beijing has already begun to diversify its trade, commercial, and industrial partners, and this process will become easier because, for both climate and economic reasons, many emerging economies and developing countries in the Global South are shifting towards renewables. At the same time, a de-concentration of low-carbon supply chains could also benefit China, as this could translate into more resilience and fewer bottlenecks for global green development. Moreover, this could also ease worries in the West about China's prominence in those supply chains and reduce the appetite for more muscular responses, which could hurt not only Beijing but also the overall global energy transition and climate agenda. Several uncertainties, however, still remain, as the Taiwan issue could escalate, which would inevitably have repercussions for the relations between China and the US and its allies in the West and the Indo-Pacific region. In this case, it is very difficult to hypothesize what the final consequences of an open great powers confrontation on the matter will be. In addition, if the climate crisis worsens—an outcome that is not very unlikely—China will feel pressure, including from the Global South, to contribute more to its solution. Its status as a developing country could be less and less sustainable from a political point of view, especially if China's economic and industrial growth continues. China itself would be very vulnerable to a mounting climate crisis, and this remains an incentive for preserving international cooperation and dialogue with the West on the global energy transition, especially because China's leadership does not extend to several areas crucial for this transition and for addressing climate change. In this regard, it is important to stress that while the energy transition will happen regardless of the presence or absence of cooperation among the three green superpowers—although in the latter case probably much more slowly—the climate crisis cannot be avoided unless this cooperation is preserved, at least at a minimum level.

Finally, it is important to stress that although this book focused on China, the EU, and the US as the current green superpowers, other countries are likely to magnify their international posture in the emerging green world order. Possible future candidates for the status of green superpower are countries with large economies, like India or Brazil, but many others might improve their position in the international system thanks to the global energy transition. In all these cases, the analytical framework developed and applied in this book can serve as a useful conceptual device to study, compare, and theorize their green foreign energy policies and green power strategies.

References

Abdelal, R. 2013. 'The profits of power: Commerce and realpolitik in Eurasia'. *Review of International Political Economy*, 20(3): pp. 421–456.

Adler, E. 1997. 'Seizing the middle ground: Constructivism in world politics'. *European Journal of International Relations*, 3(3): pp. 319–363.

Power Africa. 2017. 'Power Africa Annual Report August 2017'. Available at: https://2017-2020.usaid.gov/sites/default/files/documents/1860/PA_FINAL_508c.PDF (accessed 4 November 2023).

Agarwal, N. and L. Dupuy. 2018. 'United States and the fight against climate change: A greater role for the US EXIM Bank?'. Asia-Pacific Research and Training Network on Trade, Policy Brief No. 52, February 2018.

Aggarwal, V. K. and A. W. Reddie. 2020. 'New economic statecraft: Industrial policy in an era of strategic competition'. *Issues & Studies*, 56(02): 2040006.

Aggarwal, V. K. and A. W. Reddie. 2021. 'Economic statecraft in the 21st century: implications for the future of the global trade regime'. *World Trade Review*, 20(2): pp.137–151.

AIIB. 2021. 'Environmental and Social Framework'. Available at: https://www.aiib.org/en/policies-strategies/framework-agreements/environmental-social-framework.html (accessed 22 June 2023).

AIIB. 2022. 'Environmental and Social Framework'. Available at: https://www.aiib.org/en/policies-strategies/framework-agreements/environmental-social-framework.html (accessed 22 June 2023).

Aljazeera. 2023. 'Saudi Arabia signs $5.6bn EV deal as it eyes closer China ties'. 12 June 2023. Available at: https://www.aljazeera.com/news/2023/6/12/saudi-arabia-signs-5-6bn-ev-deal-as-it-eyes-closer-china-ties (accessed 16 July 2023).

Allen, M. M., M. L. Allen, S. I. Saqib, and J. Liu. 2021. 'State-permeated capitalism and the solar PV industry in China and India'. *New Political Economy*, 26(4): pp. 527–539.

Allison, G. 2020. 'The new spheres of influence: Sharing the globe with other great powers'. *Foreign Affairs*, 99(2): pp. 30–40.

Amsden, A. H. 2004. *Asia's Next Giant: South Korea and Late Industrialization*. New York: Oxford University Press.

Andersen, S. S., A. Goldthau and N. Sitter. 2016. 'The EU regulatory state, commission leadership and external energy governance'. In *EU Leadership in Energy and Environmental Governance*, edited by J. M. Godzimirski, pp. 51–68. London: Palgrave Macmillan.

Anderson, I. H. 1981. *Aramco, the United States, and Saudi Arabia: A study of the dynamics of foreign oil policy, 1933–1950*. Princeton: Princeton University Press.

Andrews-Speed, P. 2012. *The Governance of Energy in China. Transition to a Low-Carbon Economy*. London: Palgrave Macmillan

Andrews-Speed, P. and R. Dannreuther. (eds.). 2013. *China, Oil and Global Politics*. New York: Routledge.

Andrews-Speed, P., M. Qiu, and C. Len. 2016. 'Chinese engagement in Southeast Asian energy and mineral resources: Motivations and outlook'. *Eurasian Geography and Economics*, 57(3): pp. 316–342.

Andrews-Speed, P. and S. Zhang. 2019. *China as a Global Clean Energy Champion. Lifting the Veil*. Singapore: Palgrave Macmillan.

Ansell, C. 2000. 'The networked polity: Regional development in Western Europe'. *Governance*, 13(3): pp. 303–333.

Araujo, B. 2019. 'Regulating through trade: The contestation and recalibration of EU deep and comprehensive FTAs'. *Pace International Law Review*, 31(2): pp. 377–422.

Bach, D. and A. L. Newman. 2007. 'The European regulatory state and global public policy: micro-institutions, macro-influence'. *Journal of European Public Policy*, 14(6): pp. 827–846.

Bach, D., A. L. Newman and S. Weber. 2006. 'The international implications of China's fledgling regulatory state: From product maker to rule maker'. *New Political Economy*, 11(4): pp. 499–518.

Bäckstrand, K. and O. Elgström. 2013. 'The EU's role in climate change negotiations: From leader to "leadiator"'. *Journal of European Public Policy*, 20(10): pp. 1369–1386.

Baldwin, D. A. 1985. *Economic Statecraft*. Princeton: Princeton University Press.

Batista Barbosa, P. H. 2020. *New Kids on The Block. China's Arrival in Brazil's Electric Sector.* GCI Working Paper, n. 12, 12/2020, Boston University, Global Development Policy Center.

Batzella, F. 2018. 'Work in Progress: The Development of EU External Engagement on Energy'. In *The European Union's Evolving External Engagement: Towards New Sectoral Diplomacies?*, edited by C. Damro, S. Gstöh, and S. Schunz, pp. 107–125. London: Routledge.

Bazilian, M., M. Bradshaw, J. Gabriel, A. Goldthau, and K. Westphal. 2020. 'Four scenarios of the energy transition: Drivers, consequences, and implications for geopolitics'. *Wiley Interdisciplinary Reviews: Climate Change*, 11(2): e625.

Beringer, S. L. 2019. 'Energy, climate change and EU development policy'. In *EU Development Policies: Between Norms and Geopolitics*, edited by S. L. Beringer, S. Maier, and M. Thiel, pp. 17–34. London: Palgrave Macmillan.

Bernell, D. and C. A. Simon. 2016. *The Energy Security Dilemma. US Policy and* Practice. London: Routledge.

Bianco, C. 2023. 'Renewable relations: A strategic approach to European energy cooperation with the Gulf states'. *ECFR Policy Brief*, 16 June 2023, Rome.

Biedenkopf, K., K. Dupont, and D. Torney. 2022. 'The European Union: A Green Great Power?'. In *Great Powers, Climate Change and Global Environmental Responsibilities*, edited by R. Falkner and B. Buzan, pp. 95–116. Oxford: Oxford University Press.

Bistline, J. et al. 2023. 'Emissions and energy impacts of the Inflation Reduction Act'. *Science*, 380(6652): pp. 1324–1327.

Blackwill, R. D. and J. M. Harris. 2016. *War by Other Means: Geoeconomics and Statecraft*. Cambridge, MA: The Belknap Press of Harvard University.

Block, F. 2008. 'Swimming against the current: The rise of a hidden developmental state in the United States'. *Politics & Society*, 36(2): pp. 169–206.

Block, F. and M. R. Keller (eds.). 2011. *State of Innovation. The US Government's Role in Technology Development*. New York: Routledge.

Block, F., M. R. Keller, and M. Negoita. 2023. 'Revisiting the Hidden Developmental State'. *Politics & Society*, https://doi.org/10.1177/003232922311520

Blondeel, M., M. J. Bradshaw, G. Bridge, and C. Kuzemko. 2021. 'The geopolitics of energy system transformation: A review'. *Geography Compass*, 15(7): e12580.

BloombergNEF. 2023. 'Global Low-Carbon Energy Technology Investment Surges Past $1 Trillion for the First Time'. 26 January 2023. Available at: https://about.bnef.com/blog/global-low-carbon-energy-technology-investment-surges-past-1-trillion-for-the-first-time/ (accessed 3 August 2023).

Bocquillon, P., E. Brooks, and T. Maltby. 2023. 'Talkin' Bout a Revolution? Institutional Change in the EU Recovery and Resilience Facility: The Case of Climate Policy'. *Journal of Common Market Studies*, https://doi.org/10.1111/jcms.13536

Bodansky, D. 2001. 'The History of the Global Climate Change Regime'. In *International Relations and Global Climate Change*, edited by U. Luterbacher and D. Sprinz, pp. 23–40. Cambridge MA: MIT Press.

Bordoff, J. and M. L. O'Sullivan. 2022. 'Green Upheaval: The New Geopolitics of Energy'. *Foreign Affairs*, 101: 68.

Börzel, T. A. and T. Risse. 2012. 'From Europeanisation to Diffusion: Introduction'. *West European Politics*, 35(1): pp. 1–19.

Bradford, A. 2012. 'The Brussels effect'. *Northwestern University Law Review*, 107(1): pp. 1–68.

Bradford, A. 2020. *The Brussels effect: How the European Union rules the world.* New York: Oxford University Press.

Bretherton, C. and J. Vogler. 2005. *The European Union as a Global Actor.* London: Routledge.

Bretherton, C. and J. Vogler. 2013. 'A global actor past its peak?'. *International Relations*, 27(3): pp. 375–390.

BRICS. 2023. 'BRICS 2023. South Africa Summit, Themes and Priorities'. Available at: https://brics2023.gov.za/theme-and-priorities/ (accessed 30 August 2023).

Bulkeley, H., L. Andonova, M. M. Betsill, D. Compagnon, T. Hale, M. J. Hoffmann, P. Newell, M. Paterson, R. Charles, and S. D. Van Deveer. 2014. *Transnational Climate Change Governance.* Cambridge: Cambridge University Press.

Buzan, B. and R. Falkner. 2022. 'Great Powers and Environmental Responsibilities: A Conceptual Framework'. In *Great Powers, Climate Change and Global Environmental Responsibilities*, edited by R. Falkner and B. Buzan, pp. 14–48. Oxford: Oxford University Press.

Buzan, B. and G. Lawson. 2014. 'Capitalism and the emergent world order'. *International Affairs*, 90(1): pp. 71–91.

Buzan, B. and G. Lawson. 2015. *The Global Transformation: History, Modernity and the Making of International Relations.* Cambridge: Cambridge University Press.

Buzogány, A. 2018. 'Neighbourhood countries: Promoting environmental protection close to home'. In *European Union External Environmental Policy. Rules, Regulation and Governance Beyond Borders*, edited by C. Adelle, K. Biedenkopf, and D. Torney, pp. 233–252. London: Palgrave Macmillan.

Cambini, C. and A. Rubino (eds.). 2014. *Regional Energy Initiatives: MedReg and the Energy Community.* London: Routledge.

Carpenter, C. 2023. 'Jinko Solar celebrates adding 4 GW of solar capacity in UAE, sees Saudi Arabia as next hot spot'. *S&P Global*, 22 March 2023. Available at: https://www.spglobal.com/commodityinsights/en/market-insights/latest-news/energy-transition/032223-jinko-solar-celebrates-adding-4-gw-of-solar-capacity-in-uae-sees-saudi-arabia-as-next-hot-spot (accessed 12 April 2023).

CGEF. 2022. 'China's Global Energy Finance-Database'. Boston University Global Development Policy Center. Available at: http://www.bu.edu/cgef (accessed 4-6 March 2023).

CGEF. 2023. 'Green Horizons? China's Global Energy Finance in 2022'. Boston University Global Development Policy Center. Available at: https://www.bu.edu/gdp/2023/11/13/green-horizons-chinas-global-energy-finance-in-2022/ (accessed 11 December 2023).

CGPD. 2022. 'China's Global Power Database'. Boston University Global Development Policy Center. Available at https://www.bu.edu/cgp/ (accessed 6-9 June 2023).

Chelminski, K., L. B. Andonova, and Y. Sun. 2022. 'Emergence and structuring of the Clean Energy Regime Complex'. *Global Governance: A Review of Multilateralism and International Organizations*, 28(4): pp. 587–616.

Chen, G. 2008. 'China's diplomacy on climate change'. *The Journal of East Asian Affairs*, 22(1): pp. 145–174.

Chen, G. 2012. *China's Climate Policy.* London: Routledge.

Chen, G. and C. Lees. 2016. 'Growing China's renewables sector: A developmental state approach'. *New Political Economy*, 21(6): pp. 574–586.

Chen, X., Z. Li, K. P. Gallagher and D. L. Mauzerall. 2021. 'Financing carbon lock-in in developing countries: Bilateral financing for power generation technologies from China, Japan, and the United States'. *Applied Energy*, 300: 117318.

Cheng, W. and K. Zhang. 2023. 'Setting 'green' boundaries for Chinese green finance: Multilevel governance and regulatory stringency'. *Environmental Policy and Governance*, https://doi.org/10.1002/eet.2066

Cherp, A. and J. Jewell. 2011. 'The three perspectives on energy security: Intellectual history, disciplinary roots and the potential for integration'. *Current Opinion in Environmental Sustainability*, 3(4): pp. 202–212.

Cheung, A. B. 2008. 'The story of two administrative states: State capacity in Hong Kong and Singapore'. *The Pacific Review*, 21(2): pp. 121–145.

Christoff, P. 2010. 'Cold climate in Copenhagen: China and the United States at COP15'. *Environmental Politics*, 19(4): pp. 637–656.

Church, C. and A. Crawford. 2020. 'Minerals and the Metals for the Energy Transition: Exploring the Conflict Implications for Mineral-Rich, Fragile States'. In *The Geopolitics of the Global Energy Transition*, edited by M. Hafner and S. Tagliapietra, pp. 279–304. Cham: Springer.

Clift, B. 2014. *Comparative Political Economy: States, Markets and Global Capitalism.* London: Palgrave Macmillan.

Coenen, J., S. Bager, P. Meyfroidt, J. Newig, and E. Challies. 2021. 'Environmental governance of China's belt and road initiative'. *Environmental Policy and Governance*, 31(1): pp. 3–17.

Colgan, J. D. 2021. *Partial Hegemony. Oil Politics and International Order.* New York: Oxford University Press.

Colgan, J. D. and M. Hinthorn. 2023. 'International Energy Politics in an Age of Climate Change'. *Annual Review of Political Science*, 26: pp. 79–96.

CREA. 2022. 'Briefing: 12.8 GW of Chinese overseas coal projects cancelled, but 19 GW could still go ahead'. Centre for Research on Energy and Clean Air. Available at: https://energyandcleanair.org/wp/wp-content/uploads/2022/04/Final_Chinese-overseas-briefing_April2022.pdf (accessed 8 June 2023).

Cunningham, K., S. Dennison, S. Hix, and I. Learmonth. 2024. 'A sharp right turn: A forecast for the 2024 European Parliament elections'. ECFR Policy Brief, 23 January 2024. Available at: https://ecfr.eu/publication/a-sharp-right-turn-a-forecast-for-the-2024-european-parliament-elections/ (accessed 2 February 2024).

Cunningham, L. J. and C. M. Jordan. 2023. 'Renewable Energy and Energy Efficiency Incentives: A Summary of Federal Programs'. *Congressional Research Service Report.* Available at: https://crsreports.congress.gov/product/pdf/R/R40913 (accessed 5 September 2023).

Curran, L., P. Lv, and F. Spigarelli. 2017. 'Chinese investment in the EU renewable energy sector: Motives, synergies and policy implications'. *Energy Policy*, 101: pp. 670–682.

Currier, C. L. and M. Dorraj (eds.). 2011. *China's energy relations with the developing world.* New York: The Continuum International Publishing Group.

Damen, M. 2022. 'EU strategic autonomy 2013–2023. From concept to capacity'. Briefing EU Strategic Autonomy Monitor. Available at: https://www.europarl.europa.eu/RegData/etudes/BRIE/2022/733589/EPRS_BRI(2022)733589_EN.pdf (accessed 7 September 2023).

Damro, C. 2012. 'Market power Europe'. *Journal of European Public Policy*, 19(5): pp. 682–699.

Dannreuther, R. 2017. *Energy Security.* Cambridge: Polity Press.

Davies, S. E. and J. True. 2017. 'Norm entrepreneurship in foreign policy: William Hague and the prevention of sexual violence in conflict'. *Foreign Policy Analysis*, 13(3): pp. 701–721.

Depledge, J. 2005. 'Against the grain: The United States and the global climate change regime'. *Global Change, Peace & Security*, 17(1): pp. 11–27.

Di Carlo, D. and L. Schmitz. 2023. 'Europe first? The rise of EU industrial policy promoting and protecting the single market'. *Journal of European Public Policy*, 30(10): pp. 2063–2096.

Di Tommaso, M. R. and S. O. Schweitzer. 2013. *Industrial Policy in America: Breaking the Taboo*. Cheltenham: Edward Elgar.

Doshi, R. 2021. *The Long Game: China's Grand Strategy to Displace American Order*. Oxford: Oxford University Press.

Downie, C. 2019. *Business Battles in the US Energy Sector: Lessons for a Clean Energy Transition*. London: Routledge.

Downie, E. 2020. 'Powering the Globe: Lessons from Southeast Asia for China's Global Energy Interconnection Initiative'. Center on Global Energy Policy, Columbia University. Available at: https://www.energypolicy.columbia.edu/wp-content/uploads/2020/04/GEI_CGEP-Report_111722.pdf (accessed 7 August 2023).

Duehren, A. 2023. 'Biden Administration Explores Raising Tariffs on Chinese EVs'. *The Wall Street Journal*, 21 December 2023. Available at: https://www.wsj.com/economy/trade/biden-administration-explores-raising-tariffs-on-chinese-evs-e439d87d (accessed 6 January 2024).

Dunlap, R. E. and A. M. McCright. 2010. 'Climate Change Denial: Sources, Actors, and Strategies'. In *Routledge International Handbook of Climate Change and Society*, edited by C. Lever-Tracy, pp. 240–259. New York: Routledge.

Dupont, C., S. Oberthür, and K. Biedenkopf. 2018. 'Climate Change: Adapting to Evolving Internal and External Dynamics'. In *European Union External Environmental Policy. Rules, Regulation and Governance Beyond Borders*, edited by C. Adelle, K. Biedenkopf, and D. Torney, pp. 105–124. London: Palgrave Macmillan.

ECFR. 2023. 'EU Energy Deals Tracker'. European Council on Foreign Relations. Available at: https://ecfr.eu/special/energy-deals-tracker/?country=eu#about (accessed 27-28 October 2023).

EEA. 2021. 'Trends and projections in Europe 2021'. European Environmental Agency, Report No 13/2021, Copenhagen.

EIA. 2023. 'Oil and petroleum products explained. Oil imports and export'. US Energy Information Administration. Available at: https://www.eia.gov/energyexplained/oil-and-petroleum-products/imports-and-exports.php#:~:text=Total%20petroleum%20 nct%20cxports%20were,about%2020.01%20million%20b%2Fd (accessed 5 November 2023).

EIB. 2020. 'EIB Group Climate Bank Roadmap 2021–2025'. November 2020, Luxemburg.

EIB. 2023. 'EIB energy lending: data by country 2018–2022'. Available at: https://www.eib.org/en/projects/sectors/energy/working-together/index.htm (accessed 29 October 2023).

Eicke, L. 2023. 'US Hydrogen Policy. Paving the Way for Energy Independence, Technology Leadership and Decarbonization'. *Research Institute for Sustainability (RIFS) Discussion Paper*, December 2023, Potsdam.

Elkind, J. 2020. 'US clean energy transition and implications for geopolitics'. In *The Geopolitics of the Global Energy Transition*, edited by M. Hafner and S. Tagliapietra, pp. 47–74. Cham: Springer.

EMBER. 2023. 'China's solar exports data'. Available at: https://ember-climate.org/app/uploads/2023/09/Data-Chinas-solar-exports.xlsx (accessed 18 October 2023).

ENEL. 2023. 'Countries'. Available at: https://www.enelgreenpower.com/countries (accessed 22 December 2023).

Energy Community. 2022. 'Annual Implementation Report'. Available at: https://www.energy-community.org/implementation/report.html (accessed 3 September 2023).

Engie. 2023. 'Renewables'. Available at: https://www.engie-africa.com/node/85#:~:text=In%20Africa%2C%20ENGIE%20is%20accelerating,our%20commitment%20to%20sustainable%20energy (accessed 26 December 2023).

Escribano, G. and I. U. Arbeloa. 2023. 'Why Latin America matters to the EU on energy: Diversification, transition partners and new value chains'. Elcano Royal Institute, Madrid, 30 March 2023. Available at: https://www.realinstitutoelcano.org/en/analyses/why-latin-america-matters-to-the-eu-on-energy/ (accessed 22 January 2024).

Esarey, A., M. A. Haddad, J. I. Lewis, and S. Harrell, eds. 2020. *Greening East Asia: The Rise of the Eco-Developmental State*. Seattle, WA: University of Washington Press.

Etzioni, A. 2015. 'Spheres of influence: A reconceptualization'. *Fletcher Forum of World Affairs*, 39(2): pp. 117–132.

EU-AITF. 2020. 'Africa-EU Infrastructure Trust Fund. Operation Report 2019'. Available at: https://www.eu-africa-infrastructure-tf.net/attachments/Annual%20Reports/eu_africa_infrastructure_trust_fund_annual_report_2019_en_01.pdf (accessed 6 April 2023).

Euractive. 2023. 'China to curb exports of graphite, a key battery material'. 20 October 2023. Available at: https://www.euractiv.com/section/economy-jobs/news/china-to-curb-exports-of-graphite-a-key-battery-material/ (accessed 25 October 2023).

European Commission. 2007. *An Energy Policy for Europe*. Brussels, COM(2007) 1 final.

European Commission. 2013. *On the Implementation of the Raw Materials Initiative*. Brussels, COM(2013) 442 final.

European Commission. 2015. *A Framework Strategy for a Resilient Energy Union with a Forward-Looking Climate Change Policy*. Brussels, COM(2015) 80 final.

European Commission. 2019. *European Green Deal*. Brussels, COM(2019) 640 final.

European Commission. 2020. *A New Industrial Strategy for Europe*. Brussels, COM(2020) 102 final.

European Commission. 2021. *Building a Stronger Single Market for Europe's Recovery*. Brussels, COM(2021) 350 final.

European Commission. 2022a. 'Team Europe's contribution to global climate finance 2013–2021'. Available at: https://climate.ec.europa.eu/system/files/2022-11/climate_finance_factsheet_en.pdf (accessed 5 May 2023).

European Commission. 2022b. 'Global Gateway: First meeting of the Global Gateway Board'. Press Release, 11 December 2022. Available at: https://ec.europa.eu/commission/presscorner/detail/en/ip_22_7656 (accessed 2 August 2023).

European Commission. 2022c. *REPowerEU Plan*. Brussels, COM(2022) 230 final.

European Commission. 2022d. *An EU Strategy on Standardisation. Setting global standards in support of a resilient, green and digital EU single market*. Brussels, COM(2022) 31 final.

European Commission. 2022e. 'Budget Support. Trends and Results 2022'. September 2022, Brussels.

European Commission. 2023a. *Investment needs assessment and funding availabilities to strengthen EU's Net-Zero technology manufacturing capacity*. Brussels, SWD(2023) 68 final.

European Commission. 2023b. 'Critical Raw Materials: Ensuring secure and sustainable supply chains for EU's green and digital future'. Press release, 16 March 2023, Brussels. Available at: https://ec.europa.eu/commission/presscorner/detail/en/ip_23_1661 (accessed 6 August 2023).

European Commission. 2023c. 'Joint Statement from the European Union and the United States'. Press release, 9 September 2023. Available at: https://ec.europa.eu/commission/presscorner/detail/en/statement_23_4419 (accessed 28 December 2023).

European Commission. 2023d. 'Main Outcomes of the Mapping of External Financial Tools of the EU'. Joint Staff Working Document, April 2023, Brussels.

European Commission-High Representative of the Union for Foreign Affairs and Security Policy. 2023. 'European Economic Security Strategy'. Brussels, JOIN(2023) 20 final.

European Commission-High Representative of the Union for Foreign Affairs and Security Policy 2021. *The Global Gateway*. Brussels, JOIN(2021) 30 final.

European Council. 2021. 'Council conclusions on Climate and Energy Diplomacy. Delivering on the external dimension of the European Green Deal'. 25 January 2021. Available at: https://www.consilium.europa.eu/media/48057/st05263-en21.pdf (accessed 24 August 2023).

European Parliament. 2023. 'EU anti-subsidy probe into electric vehicle imports from China'. Available at: https://www.europarl.europa.eu/RegData/etudes/ATAG/2023/754553/EPRS_ATA(2023)754553_EN.pdf (accessed 3 January 2024).

Eurostat. 2023. 'EU trade with Russia: Latest developments'. May 2023. Available at: https://ec.europa.eu/eurostat/statistics-explained/index.php?title=EU_trade_with_Russia_-_latest_developments#Key_product_groups_imported_by_the_EU_from_Russia (accessed 5 June 2023).

Ewing, J., M. Ross, A. Pickle, R. Stout, and B. Murray. 2022. 'Pathways to Net-Zero for the US Energy Transition'. Nicholas Institute for Energy, Environment & Sustainability, Duke University. Available at: https://dukespace.lib.duke.edu/items/ce473216-aa37-4ab3-8af0-906ee97dfbd3 (accessed 4 October 2023).

EXIM. 2009. 'EXIM Bank Annual Report 2009'. Available at: https://www.exim.gov/news/reports (accessed 8 June 2023).

EXIM. 2016. 'EXIM Bank Annual Report 2016'. Available at: https://www.exim.gov/news/reports (accessed 8 June 2023).

Farrell, H. and A. Newman. 2019. 'Weaponized interdependence: How global economic networks shape state coercion'. *International Security*, 44(1): pp. 42–79.

Feng, T. T., X. L. Gong, Y. H. Guo, Y. S. Yang, B. B. Pan, S. P. Li, and J. Dong. 2020. 'Electricity cooperation strategy between China and ASEAN countries under "The Belt and road"'. *Energy Strategy Reviews*, 30, 100512.

Finamore, B. 2018. *Will China Save the Planet?*. Cambridge: Polity Press.

Finnemore, M. and K. Sikkink. 1998. 'International norm dynamics and political change'. *International Organization*, 52(4): pp. 887–917.

Fischhendler, I., L. Herman, and N. Maoz. 2017. 'The political economy of energy sanctions: Insights from a global outlook 1938–2017'. *Energy Research & Social Science*, 34, pp. 62–71.

Frankfurt School-UNEP Centre/BNEF. 2020. 'Global Trends in Renewable Energy Investment 2020'. Available at: https://www.fs-unep-centre.org/wp-content/uploads/2020/06/GTR_2020.pdf (accessed 30 June 2023).

Gallagher, K. S. and Q. Qi. 2018. 'Policies governing China's overseas development finance implications for climate change'. Center for International Environment and Resource Policy, The Fletcher School, Tufts University, March, 16. Available at: https://sites.tufts.edu/cierp/files/2018/03/CPL_ChinaOverseasDev.pdf (accessed 22 July 2024).

Gang, C. 2019. *Politics of Renewable Energy in China*. Cheltenham: Edward Elgar.

Genç, H. O. and K. A. Akkemik. 2018. 'Developmental State, Industrial Policy, and Green Growth in China'. In *Industrial Policy and Sustainable Growth*, edited by M. A. Yülek, pp. 285–310. Singapore: Springer.

Geri, L. R. and D. E. McNabb. 2017. *Energy Policy in the U.S. Politics, Challenges, and Prospects for Change*. London: Routledge.

Gertz, G. and M. M. Evers. 2020. 'Geoeconomic competition: Will state capitalism win?' *The Washington Quarterly*, 43(2): pp. 117–136.

Gilpin, R. 2011. *Global Political Economy*. Princeton: Princeton University Press.

Goel, R. 2004. 'A bargain born of a paradox: The oil industry's role in american domestic and foreign policy'. *New Political Economy*, 9(4): pp. 467–492.

Goldthau, A. and L. Hughes. 2020. 'Protect global supply chains for low-carbon technologies'. *Nature*, 585: pp. 28–30.

Goldthau, A. and N. Sitter. 2014. 'A liberal actor in a realist world? The Commission and the external dimension of the single market for energy'. *Journal of European Public Policy*, 21(10): pp. 1452–1472.

Goldthau, A. and N. Sitter. 2015a. *A Liberal Actor in a Realist World: The European Union Regulator State and the Global Political Economy of Energy*. Oxford: Oxford University Press.

Goldthau, A. and N. Sitter. 2015b. 'Soft power with a hard edge: EU policy tools and energy security'. *Review of International Political Economy*, 22(5): pp. 941–965.

Goldthau, A. and N. Sitter. 2018. 'Regulatory or Market Power Europe? EU Leadership Models for International Energy Governance'. In *New Political Economy of Energy in Europe. Power to Project, Power to Adapt*, edited by J. M. Godzimirski, pp. 27–47. London: Palgrave Macmillan.

Goldthau, A. and R. Youngs. 2023. 'The EU Energy Crisis and a New Geopolitics of Climate Transition'. *Journal of Common Market Studies*, https://doi.org/10.1111/jcms.13539

Gong, X., A. Boute, and R. Quitzow. 2023. 'China's Emerging Hydrogen Economy: Policies, Institutions, Actors'. RIFS Study, January 2023, Research Institute for Sustainability, Potsdam.

Van de Graaf, T. 2013. 'Fragmentation in Global Energy Governance: Explaining the Creation of IRENA'. *Global Environmental Politics*, 13(3): pp. 14–33.

Van de Graaf, T. and J. D. Colgan. 2017. 'Russian gas games or well-oiled conflict? Energy security and the 2014 Ukraine crisis'. *Energy Research & Social Science*, 24: pp. 59–64.

Van de Graaf, T., I. Overland, D. Scholten, and K. Westphal. 2020. 'The new oil? The geopolitics and international governance of hydrogen'. *Energy Research & Social Science*, 70, 101667.

Van de Graaf, T. and B. K. Sovacool. 2020. *Global Energy Politics*. Cambridge: Polity Press.

Van de Graaf, T., B. K. Sovacool, A. Ghosh, F. Kern, and M. T. Klare (eds.). 2016. *The Palgrave Handbook of the International Political Economy of Energy*. London: Palgrave Macmillan.

Grohol, M. and C. Veeh. 2023. 'Study on the Critical Raw Materials for the EU. Final Report'. Available at: https://op.europa.eu/en/publication-detail/-/publication/57318397-fdd4-11ed-a05c-01aa75ed71a1 (accessed 4 September 2023).

Grünberg, N. 2017. 'Revisiting Fragmented Authoritarianism in China's Central Energy Administration'. In *Chinese Politics as Fragmented Authoritarianism. Earthquakes, Energy and Environment*, edited by K. E. Brodsgaard, pp. 15–37. London: Routledge.

Grünberg, N. and K. Drinhausen. 2019. 'The Party Leads on Everything: China's changing governance in Xi Jinping's new era'. *Merics China Monitor*, 24 September 2019. Available at: https://merics.org/en/report/party-leads-everything (accessed 22 July 2024).

Gulley, A. L., E. A. McCullough, and K. B. Shedd. 2019. 'China's domestic and foreign influence in the global cobalt supply chain'. *Resources Policy*, 62: pp. 317–323.

Gürtler, K., R. Postpischil, and R. Quitzow. 2019. 'The dismantling of renewable energy policies: The cases of Spain and the Czech Republic'. *Energy Policy*, 133, 110881.

Gustavsson, J. 1999. 'How should we study foreign policy change?'. *Cooperation and Conflict*, 34(1): pp. 73–95.

GWEC. 2022. *Global Wind Report 2022*. Global Wind Energy Council, Brussels.

Hafner, M. and S. Tagliapietra (eds.). 2020. *The Geopolitics of the Global Energy Transition*. Cham: Springer.

Hafner, M. and A. Wochner. 2020. 'How tectonic shift in global energy are affecting global governance'. In *Global Governance in Transformation*, edited by L. Grigoryev and A. Pabst, pp. 147–162. Cham: Springer.

Hall, P. A. 1993. 'Policy paradigms, social learning, and the state: The case of economic policymaking in Britain'. *Comparative Politics*, 25: pp. 275–296.

Haroche, P. 2023. 'A 'geopolitical Commission': Supranationalism meets global power competition'. *Journal of Common Market Studies*, 61(4): pp. 970–987

Heine, J. 2013. 'From club to network diplomacy'. In *The Oxford Handbook of Modern Diplomacy*, edited by A. F. Cooper, pp. 54–69. Oxford: Oxford University Press.

Henares, T. J. and L. L. Delina. 2022. 'Beyond the narratives of trade and domination: How ecology, connectivity, and history shape Chinese investments in foreign electricity grids'. *Energy Research & Social Science*, 93, 102823.

Hermann, C. F. 1990. 'Changing course: When governments choose to redirect foreign policy'. *International Studies Quarterly*, 34(1): pp. 3–21.

Herranz-Surralles, A. 2024. 'The EU energy transition in a geopoliticizing world'. *Geopolitics*, https://doi.org/10.1080/14650045.2023.2283489

Herranz-Surrallés, A. 2015. 'European External Energy Policy: Governance, Diplomacy and Sustainability'. In *Sage Handbook of European Foreign Policy*, edited by A. K. Aarstad, E. Drieskens, K. E. Jørgensen, K. Laatikainen, and B. Tonra (eds.), pp. 913–938. London: Sage.

Herranz-Surrallés, A. 2016. 'An emerging EU energy diplomacy? Discursive shifts, enduring practices'. *Journal of European Public Policy*, 23(9): pp. 1386–1405.

Herranz-Surrallés, A. 2018. 'Thinking energy outside the frame? Reframing and misframing in Euro-Mediterranean energy relations'. *Mediterranean Politics*, 23(1): pp. 122–141.

Herrero, A. G. 2021. 'What is behind China's dual circulation strategy'. *China Leadership Monitor, Fall 2021, Issue 69*: pp. 1–12.

Heung, P., J. Rydge, and W-S. Chan. 2023. 'CBAM: Country and sector impacts start to emerge'. HSBC Global Research, 18 October 2023. Available at: https://www.research.hsbc.com/C/1/1/320/lbwqfVG (accessed 12 December 2023).

Hildingsson R., J. Stripple, and A. Jordan. 2012. 'Governing renewable energy in the EU: Confronting a governance dilemma'. *European Political Science*, 11: pp. 18–30.

Hilton, I. and O. Kerr. 2017. 'The Paris Agreement: China's "new normal" role in international climate negotiations'. *Climate Policy*, 17(1): pp. 48–58.

Hocking, B. 1999. 'Catalytic Diplomacy: Beyond Newness and Decline'. In *Innovation in Diplomatic Practice*, edited by J. Melissen, pp. 97–116. London: Palgrave Macmillan.

Hocking, B. 2004. 'Diplomacy'. In *Contemporary European Foreign Policy*, edited by W. Carlsnaes, H. Sjursen and B. White, pp. 91–110. London: Sage.

Hoffmann, C. and C. Ergenc. 2023. 'A greening dragon in the desert? China's role in the geopolitical ecology of decarbonisation in the Eastern Mediterranean'. *Journal of Balkan and Near Eastern Studies*, 25(1): pp. 82–101.

Högselius, P. 2013. *Red Gas: Russia and the Origins of European Energy Dependence*. London: Palgrave Macmillan.

Home, A. 2023. 'US looks to shut China out of its battery supply chain'. *Reuters*, 6 December 2023. Available at: https://www.reuters.com/markets/commodities/us-looks-shut-china-out-its-battery-supply-chain-2023-12-05/ (accessed 8 January 2024).

Hopewell, K. 2021. 'Power transitions and global trade governance: The impact of a rising China on the export credit regime'. *Regulation & Governance*, 15(3): pp. 634–652.

Hopkins, M. and Y. Li. 2016. 'The Rise of the Chinese Solar Photovoltaic Industry: Firms, Governments, and Global Competition'. In *China as an Innovation Nation*, edited by Y. Zhou, W. Lazonick, and Y. Sun, pp. 306–332. Oxford: Oxford University Press.

Hudson, V. M. and B. S. Day. 2007. *Foreign Policy Analysis: Classic and Contemporary Theory*. London: Rowman & Littlefield.

Hughes, L. and J. Meckling. 2017. 'The politics of renewable energy trade: The US-China solar dispute'. *Energy Policy*, 105: pp. 256–262.

Hughes, L. and R. Quitzow. 2018. 'Low-Carbon Technologies, National Innovation Systems, and Global Production Networks: The State of Play'. In *Handbook of the International Political Economy of Energy and Natural Resources*, edited by A. Goldthau, M. F. Keating, and C. Kuzemko, pp. 281–296. Cheltenham: Edward Elgar.

Hulbert, M. and A. Goldthau. 2013. 'Natural Gas Going Global? Potentials and Pitfalls'. In *The Handbook of Global Energy Policy*, edited by A. Goldthau, pp. 98–112: West Sussex: Wiley-Blackwell.

ICBC. 2018. 'Belt and Road Interim Report. Tracking evolving scope, discovering expanding opportunities'. *ICBC Standard*, April 2018. Available at: https://v.icbc.com.cn/userfiles/Resources/ICBC/haiwai/StandardBank/Download/2019/InauguralWhitepapers3.pdf (accessed 4 July 2023).

IEA. 2016. *Boosting the Power Sector in Sub-Saharan Africa*. Paris: International Energy Agency.

IEA. 2021a. *Net Zero by 2050. A Roadmap for the Global Energy Sector*. Paris: International Energy Agency.

IEA. 2021b. *World Energy Outlook 2021*. Paris: International Energy Agency.

IEA. 2021c. *The Role of Critical Minerals in Clean Energy Transition*. Paris: International Energy Agency.

IEA. 2021d. *Patents and the energy transition. Global trends in clean energy technology innovation*. Paris: International Energy Agency.

IEA. 2022a. *Securing Clean Energy Technology Supply Chains*. Paris: International Energy Agency.

IEA. 2022b. *World Energy Investment 2022*. Paris: International Energy Agency.

IEA. 2022c. *Tracking Clean Energy Innovation. Focus on China*. Paris: International Energy Agency.

IEA. 2023a. *Energy Technologies Perspectives 2023*. Paris: International Energy Agency.

IEA. 2023b. *The State of Clean Technology Manufacturing. An Energy Technology Perspectives Special Briefing*. Paris: International Energy Agency.

IEA. 2023c. *World Energy Outlook 2023*. Paris: International Energy Agency.

IEA. 2024. *Renewables 2023*. Paris: International Energy Agency.

IISD. 2021. 'The Energy Policy Tracker: China'. International Institute for Sustainable Development. Available at: https://www.energypolicytracker.org/country/china/ (accessed 15 June 2023).

IISD. 2022. *Navigating Energy Transitions. Mapping the road to 1.5°C*. Winnipeg, Canada: International Institute for Sustainable Development.

Ikenberry, G. J. 1988. 'Market solutions for state problems: The international and domestic politics of American oil decontrol'. *International Organization*, 42(1): pp. 151–177.

IPCC. 2022. 'Sixth assessment report. Working group III: Mitigation of climate change'. Intergovernmental Panel on Climate Change. Available at: https://www.ipcc.ch/report/ar6/wg3/ (accessed 15 September 2023).

IRENA. 2018. 'Renewable capacity statistics 2018'. Available at: https://www.irena.org/publications/2018/mar/renewable-capacity-statistics-2018 (accessed 12 May 2023)

IRENA. 2019. 'A New World: The Geopolitics of the Energy Transformation'. International Renewable Energy Agency. Available at: https://www.irena.org/publications/2019/Jan/A-New-World-The-Geopolitics-of-the-Energy-Transformation (accessed 25 November 2023).

IRENA. 2022a. 'Geopolitics of the Energy Transformation. The Hydrogen Factor'. International Renewable Energy Agency. Available at: https://www.irena.org/Publications/2022/Jan/Geopolitics-of-the-Energy-Transformation-Hydrogen (accessed 22 December 2022).

IRENA. 2022b. 'Renewable energy patents data'. International Renewable Energy Agency. Available at: https://wwwirenaorg/Data/View-data-by-topic/Innovation-and-Technology/Patents-Evolution, accessed 10-14 November 2022.

IRENA. 2022c. 'Renewable Power Generation Costs in 2021'. International Renewable Energy Agency. Available at: https://www.irena.org/publications/2022/Jul/Renewable-Power-Generation-Costs-in-2021 (accessed 22 September 2023).

IRENA. 2023a. 'Renewable capacity statistics 2023'. International Renewable Energy Agency. Available at: https://www.irena.org/Publications/2023/Mar/Renewable-capacity-statistics-2023 (accessed 18 August 2023).

IRENA. 2023b. 'Geopolitics of the Energy Transition: Critical Materials'. International Renewable Energy Agency. Available at: https://www.irena.org/Publications/2023/Jul/Geopolitics-of-the-Energy-Transition-Critical-Materials (accessed 15 October 2023).

Jack, A. W., D. Luchsinger, D. B. Levine, and J. Koo. 2023. 'The Biden Administration Unveils the Long-Waited Guidance on "Foreign Entity of Concern"'. Available at: https://www.insideenergyandenvironment.com/2023/12/the-biden-administration-unveils-the-long-waited-guidance-on-foreign-entity-of-concern/ (accessed 7 January 2024).

Jaffe, A. M. 2023. 'US Defense Strategy: Forging an Industrial Orientation Towards Energy Security and Foreign Policy'. In *Handbook on the Geopolitics of the Energy Transition*, edited by D. Scholten, pp. 388–403. Cheltenham: Edward Elgar.

Jakob, M. and L. Martini. 2023. 'How can the G7 and G20 improve Just Energy Transition Partnerships? Taking stock of Just Energy Transition Partnerships'. *Ecological Institute Briefing*, June 2023. Available at: https://www.wwf.de/fileadmin/fm-wwf/Publikationen-PDF/Klima/Policy-Brief-JETPs.pdf (accessed 26 October 2023).

Jordan, A., D Huitema, H. van Asselt, T. Rayner, and F. Berkhout (eds.). 2010. *Climate change policy in the European Union*. Cambridge: Cambridge University Press.

Joskow, P. L. 2001. 'US Energy Policy during the 1990s'. National Bureau of Economic Research Working Paper 8454, July 2001.

Jotzo, F., J. Depledge, and H. Winkler. 2018. 'US and international climate policy under President Trump'. *Climate Policy*, 18(7): pp. 813–817.

Kalantzakos, S. 2017a. *China and the Geopolitics of Rare Earths*. Oxford: Oxford University Press.

Kalantzakos, S. 2017b. *The EU, US and China Tackling Climate Change. Policies and Alliances for the Anthropocene*. London: Routledge.

Kalantzakos, S. 2020. 'The race for critical minerals in an era of geopolitical realignments'. *The International Spectator*, 55(3): pp. 1–16.

Kearn, D. W. 2011. 'The hard truths about soft power'. *Journal of Political Power*, 4(1): pp. 65–85.

Keller, M. R. and F. Block. 2015. 'Do as I say or as I do? US innovation and industrial policy since the 1980s'. In *Development and Modern Industrial Policy in Practice: Issues and Country Experiences*, edited by J. Felipe, pp. 219–246. Cheltenham: Edward Elgar.

Keohane, R. O. 1982. 'State power and industry influence: American foreign oil policy in the 1940s'. *International Organization*, 36(1): pp. 165–183.

Keohane, R. O. and J. S. Nye. 1977. *Power and Interdependence*. Boston: Little Brown.

Keohane, R. O. and D. G. Victor. 2011. 'The regime complex for climate change'. *Perspectives on Politics*, 9(1): pp. 7–23.

Kleimann, D., N. Poitiers, A. Sapir, S. Tagliapietra, N. Véron, R. Veugelers, and J. Zettelmeyer. 2023. 'Green tech race? The US Inflation Reduction Act and the EU Net Zero Industry Act'. *The World Economy*, 46(12): pp. 3420–3434.

Knodt, M., M. Ringel, and R. Müller. 2020. '"Harder" soft governance in the European Energy Union'. *Journal of Environmental Policy & Planning*, 22(6): pp. 787–800.

Kong, B. 2009. *China's International Petroleum Policy*. Santa Barbara, CA: Praeger.

Kong, B. and K. P. Gallagher. 2020. 'Chinese development finance for solar and wind power abroad'. *GCI Working Paper* No. 9/2020, Global Development Policy Center, Boston University.

Kong, B. and K. P. Gallagher. 2021. 'Inadequate demand and reluctant supply: The limits of Chinese official development finance for foreign renewable power'. *Energy Research & Social Science*, 71, 101838.

Konisky, D. M. and N. D. Woods. 2016. 'Environmental policy, federalism, and the Obama presidency'. *Publius: The Journal of Federalism*, 46(3): pp. 366–391.

Kornes, M. 2020. *Wind and Solar Energy Transition in China*. London: Routledge.

Kozloff, K. and O. Shobowale. 1994. *Rethinking Development Assistance for Renewable Electricity*. Washington, DC: World Resources Institute.

Krasner, S. D. 1988. 'Sovereignty: An Institutional Perspective'. *Comparative Political Studies*, 21(1): pp. 66–94.

Kratz, A., M. J. Zenglein, G. Sebastian, and M. Witzke. 2023. 'EV Battery Investments Cushion Drop to Decade Low Chinese FDI in Europe'. *MERICS Report*. Available at: https://merics.org/sites/default/files/2023-05/merics-rhodium-group-chinese-fdi-in-europe-2022%20%281%29.pdf (accessed 14 July 2023).

Kuteleva, A. 2021. *China's Energy Security and Relations with Petrostates*. London: Routledge.

Kuzemko, C., A. Lawrence, and M. Watson. 2019. 'New directions in the international political economy of energy'. *Review of International Political Economy*, 26(1): pp. 1–24.

Ladislaw, S. 2017. 'Dissecting the idea of US energy dominance'. *Oxford Energy Forum*, Issue 111. Available at: https://www.oxfordenergy.org/wpcms/wp-content/uploads/2018/01/OEF-111.pdf (accessed 5 December 2023).

Ladislaw, S., E. Zindler, J. Nakano, N. Tsafos, L. Carey, L. Goldie-Scot, P. Lezcano, and J. Chase. 2021.*Industrial Policy, Trade, and Clean Energy Supply Chains*. Center for Strategic and International Studies. Available at: https://www.csis.org/analysis/industrial-policy-trade-and-clean-energy-supply-chains (accessed 4 April 2023).

Laird, F. N. and C. Stefes. 2009. 'The diverging paths of German and United States policies for renewable energy: Sources of difference'. *Energy Policy*, 37(7): pp. 2619–2629.

Lavenex, S. 2011. 'Concentric circles of flexible 'EUropean'integration: A typology of EU external governance relations'. *Comparative European Politics*, 9: pp. 372–393.

Lavenex, S. 2014. 'The power of functionalist extension: How EU rules travel'. *Journal of European Public Policy*, 21(6): pp. 885–903.

Lavenex, S., I. Krizic, and O. Serrano. 2017. 'EU and US regulatory power under strain? Emerging countries and the limits of external governance'. *European Foreign Affairs Review*, 22: pp. 1–18.

Lavenex, S., O, Serrano, and T. Büthe. 2021. 'Power transitions and the rise of the regulatory state: Global market governance in flux'. *Regulation & Governance*, 15(3): pp. 445–471.

Lawson, M. L. 2017. 'Major Foreign Aid Initiatives Under the Obama Administration: A Wrap-Up'. *Congressional Research Service Report*. Available at: https://crsreports.congress.gov/product/pdf/R/R44727 (accessed 3 December 2023).

Leal-Arcas, R. and S. Minas. 2016. 'Mapping out the international and European governance of renewable energy'. *Yearbook of European Law*, 35(1): pp. 621–666.

Leonard, M. 2016. *Connectivity Wars: Why Migration, Finance and Trade are the Geo-economic Battlegrounds of the Future*. London: European Council on Foreign Relations.

Leoni, Z. 2022. 'The economy-security conundrum in American grand strategy: Foreign economic policy toward China from Obama to Biden'. *China International Strategy Review*, 4(2): pp. 320–334.

Lepont, U. and M. Thiemann. 2024. 'The European Investor State: Its characteristics, genesis, and effects'. *Competition & Change*, https://doi.org/10.1177/10245294231215481

Leruth, L., P. Regibeau, L. Renneboog, and A. Mazarei. 2022. 'Green Energy Depends on Critical Minerals. Who Controls the Supply Chains?'. *ECGI Working Paper* No. 846/2022, Washington, DC.

Lesage, D., T. Van de Graaf, and K. Westphal. 2010. *Global Energy Governance in a Multipolar World*. Burlington: Ashgate.

Levi-Faur, D. 2005. 'The global diffusion of regulatory capitalism'. *The Annals of the American Academy*, 598(1): pp. 12–32.

Lewis, J. I. 2014. 'The rise of renewable energy protectionism: Emerging trade conflicts and implications for low carbon development'. *Global Environmental Politics*, 14(4): pp. 10–35.

Lewis, J. I. 2017. 'Green Energy Innovation in China'. In *Routledge Handbook of Environmental Policy in China*, edited by E. Sternfeld, pp. 280–290. London: Routledge.

Lewis, J. I. 2023a. *Cooperating for the Climate. Learning from International Partnerships in China's Clean Energy Sector*. Cambridge, MA: MIT Press.

Lewis J. I. 2023b. 'Database of Bilateral Clean Energy and Climate Agreements with China'. Mendeley Data, V1, doi: 10.17632/8rztp5ng86.1. Available at: https://data.mendeley.com/datasets/8rztp5ng86/1 (accessed 16–20 May 2023).

Li, Y., H. Cheng, R. J. S. Beeton, T. Sigler, and A. Halog. 2016. 'Sustainability from a Chinese cultural perspective: The implications of harmonious development in environmental management'. *Environment, Development and Sustainability*, 18(3): pp. 679–696.

Li, Y. and J. Shapiro. 2020. *China Goes Green: Coercive Environmentalism for a Troubled Planet*. Cambridge: Polity Press.

Liao, J. C. 2021. 'China's Green Mercantilism and Environmental Governance: A New Belt and Road to the Global South?'. In *Essays on the Rise of China and Its Implications*, edited by A. M. Denmark and L. Myers, pp. 242–269. Washington, DC: Wilson Center.

Liao, J. C. 2022. 'Talking green, building brown: China-ASEAN environmental and energy cooperation in the BRI era'. *Asian Perspective*, 46(1): pp. 21–47.

Lieberthal, K. G. and M. Oksenberg. 1988. *Policy Making in China. Leaders, Structures and Processes*. Princeton: Princeton University Press.

Limenta, M. 2023. 'Shaping trade in goods relevant to renewable energy generation: The RCEP's potential and limitation'. *Asia Pacific Law Review*, 31(2): pp. 469–488.

Lin, H. 2019. *Energy Policies and Climate Change in China. Actors, Implementation, and Future Prospects*. London: Routledge.

Lind, M. 1992. 'The Catalytic State'. *The National Interest*, 27: pp. 3–12.

Lowi, T. J. 1967. 'Making Democracy Safe for the World: National Politics and Foreign Policy'. In *Domestic Source of Foreign Policy*, edited by J. N. Rosenau, pp. 295–331. New York: The Free Press.

Luttwak, E. N. 1990. 'From Geopolitics to Geo-Economics: Logic of Conflict, Grammar of Commerce'. *The National Interest*, 20: pp. 17–23.

Ma, T. 2017. 'China upgrades climate aid to the Global South'. *China Dialogue*, 20 September 2017. Available at: https://chinadialogue.net/en/business/10086-china-upgrades-climate-aid-to-the-global-south/ (accessed 4 April 2023).

Ma, X., C. Han Springer, and H. Shao. 2022. 'Outlier or New Normal? Trends in China's Global Energy Finance'. Global Development Policy Center, Policy Brief 011, March 2022.

Available at: https://www.bu.edu/gdp/files/2022/03/GCI_PB_011_FIN.pdf (accessed 10 February 2023).

Macikenaite, V. 2020. 'China's economic statecraft: The use of economic power in an interdependent world'. *Journal of Contemporary East Asia Studies*, 9(2): pp. 108–126.

MacNeil, R. 2013. 'Seeding an energy technology revolution in the United States: Re-conceptualising the nature of innovation in liberal-market economies'. *New Political Economy*, 18(1): pp. 64–88.

MacNeil, R. 2014. 'Climate Policy, Energy Technologies, and the American Developmental State'. In *Climate Innovation. Liberal Capitalism and Climate Change*, edited by N. E. Harrison and J. Mikler, pp. 45–71. London: Palgrave Macmillan.

MacNeil, R. 2016. 'Between innovation and industrial policy: How Washington succeeds and fails at renewable energy'. *Prometheus*, 34(3–4): pp. 173–189.

MacNeil, R. and M. Paterson. 2012. 'Neoliberal climate policy: From market fetishism to the developmental state'. *Environmental Politics*, 21(2): pp. 230–247.

Mah, L. 2018. 'Promoting private sector for development: the rise of blended finance in EU aid architecture'. Working Paper CESA No. 171/2018.

Majone, G. 1997. 'From the positive to the regulatory state: Causes and consequences of changes in the mode of governance'. *Journal of Public Policy*, 17(2): pp. 139–167.

Manners, I. 2013. 'Assessing the decennial, reassessing the global: Understanding European Union normative power in global politics'. *Cooperation and Conflict*, 48(2): pp. 304–329.

Mazzucato, M. 2015. *The Entrepreneurial State: Debunking Public vs. Private Sector Myths*. London: Anthem Press.

McCarthy, D. and M. V. Olano. 2023. 'The US climate law is fueling a factory frenzy'. Canary Media, 16 August 2023. Available at: https://www.canarymedia.com/articles/clean-energy-manufacturing/the-us-climate-law-is-fueling-a-factory-frenzy-heres-the-latest-tally?utm_campaign=canary-social&utm_source=twitter&utm_medium=social&utm_content=1692191013 (accessed 22 November 2023).

McCright, A. M. and R. E. Dunlap. 2011. 'The politicization of climate change and polarization in the American public's views of global warming, 2001–2010'. *The Sociological Quarterly*, 52(2): pp. 155–194.

McFarland, V. and J. D. Colgan. 2023. 'Oil and power: The effectiveness of state threats on markets'. *Review of International Political Economy*, 30(2): pp. 487–510

McGee, J. 2014. 'The Influence of US Neoliberalism on International Climate Change Policy'. In *Climate Innovation. Liberal Capitalism and Climate Change*, edited by N. E. Harrison and J. Mikler, pp. 193–214. London: Palgrave Macmillan.

Meckling, J. 2018. 'The developmental state in global regulation: Economic change and climate policy'. *European Journal of International Relations*, 24(1): pp. 58–81.

Meckling, J. and T. Van de Graaf. 2020. 'To Engage, or Not to Engage? Us Foreign Policy on Climate and Clean Energy'. In *Handbook of US Environmental Policy*, edited by D. M. Konisky, pp. 302–315. Cheltenham: Edward Elgar.

Meckling, J., B. Kong, and T. Madan. 2015. 'Oil and state capitalism: Government-firm coopetition in China and India'. *Review of International Political Economy*, 22(6): pp. 1159–1187.

Meidan, M. 2020. 'China: Climate Leader and Villain'. In *The Geopolitics of the Global Energy Transition*, edited by M. Hafner and S. Tagliapietra, pp. 75–90. Cham: Springer.

Meidan, M., P. Andrews-Speed, A. Hove, and Y. Qin. 2023. 'China's Climate and Energy Policy After the Two Sessions: More Wait and See'. *OIES Energy Comment*, April 2023. Oxford: Oxford Institute for Energy Studies. Available at: https://www.oxfordenergy.org/publications/chinas-climate-and-energy-policy-after-the-two-sessions-more-wait-and-see/ (accessed 23 July 2024).

Meidan, M., A. Hove, and P. Andrews-Speed. 2023. 'China Re-opens: Implications for Energy Markets and Policies'. *OIES Energy Comment*, February 2023. Oxford: Oxford Institute for Energy Studies. Available at: https://www.oxfordenergy.org/wpcms/wp-content/uploads/2023/02/China-re-opens-implications-for-energy-markets-and-policies.pdf (accessed 25 July 2024).

Mertens, D. and M. Thiemann. 2018. 'Market-based but state-led: The role of public development banks in shaping market-based finance in the European Union'. *Competition & Change*, 22(2): 184–204.

Mertens, D. and M. Thiemann. 2019. 'Building a hidden investment state? The European investment bank, national developments banks and European economic governance'. *Journal of European Public Policy*, 26(1): pp. 23–43.

Meunier, S. and K. Nicolaidis. 2019. 'The geopoliticization of European trade and investment Policy'. *Journal of Common Market Studies*, 57(1): pp. 103–113.

Minas, S. and V. Ntousas. 2019. *EU Climate Diplomacy. Politics, Law and Negotiations.* London: Routledge.

Mingjiang, L. 2008. 'China debates soft power'. *The Chinese Journal of International Politics*, 2(2): pp. 287–308.

MIT. 2011. *The Future of Natural Gas: An Interdisciplinary MIT Study.* Boston: Massachusetts Institute of Technology Energy Initiative.

Mocanu, D. and M. Thiemann. 2023. 'Breeding "unicorns": Tracing the rise of the European investor state in the European venture capital market'. *Competition & Change*, https://doi.org/10.1177/1024529423120498

Moe, E. 2015. *Renewable Energy Transformation or Fossil Fuel Backlash: Vested Interests in the Political Economy.* London: Palgrave Macmillan.

Morgenstern, E. M., N. Cook, N. M. Brown, T. Lum, L. Blanchfield, P. J. Meyer, and S. I. Akhtar. 2021. 'Selected Trump Administration Foreign Aid Priorities: A Wrap-Up'. *Congressional Research Service Report.* Available at: https://crsreports.congress.gov/product/pdf/R/R46656 (accessed 9 July 2023).

Nahm, J. 2021. *Collaborative Advantage: Forging Green Industries in the New Global Economy.* Oxford: Oxford University Press.

NCSL. 2021. 'State Renewable Portfolio Standards and Goals'. National Conference of State Legislatures, 13 August 2021. Available at: https://www.ncsl.org/energy/state-renewable-portfolio-standards-and-goals (accessed 5 December 2023).

NDB. 2022. 'New Development Bank General Strategy for 2022–2026'. Available at: https://www.ndb.int/wp-content/uploads/2022/07/NDB_StrategyDocument_Eversion-1.pdf (accessed 16 July 2023).

Nedopil, C. 2022. 'Green finance for soft power: An analysis of China's green policy signals and investments in the Belt and Road Initiative'. *Environmental Policy and Governance*, 32(2): pp. 85–97.

Nedopil, C. 2023a. 'China Belt and Road Initiative Investment Report 2022'. Green Finance & Development Center, FISF Fudan University, Shanghai.

Nedopil, C. 2023b. 'China Belt and Road Initiative Investment Report 2023 H1'. Available at: https://greenfdc.org/china-belt-and-road-initiative-bri-investment-report-2023-h1/?cookie-state-change=1706714578567 (accessed 22 December 2023).

Newnham, R. 2011. 'Oil, carrots, and sticks: Russia's energy resources as a foreign policy tool'. *Journal of Eurasian Studies*, 2(2): pp. 134–143.

Nielsen, K. L. and M. Vilson. 2014. 'The Eastern partnership: Soft power strategy or policy failure?'. *European Foreign Affairs Review*, 19(2): pp. 243–262.

Norris, W. J. 2018. *Chinese Economic Statecraft: Commercial Actors, Grand Strategy, and State Control.* Ithaca, NY: Cornell University Press.

NSS. 2017. 'National Security Strategy of the United States of America'. Available at: https://trumpwhitehouse.archives.gov/wp-content/uploads/2017/12/NSS-Final-12-18-2017-0905.pdf (accessed 8 November 2023).

NSS. 2022. 'US National Security Strategy'. Available at: https://www.whitehouse.gov/wp-content/uploads/2022/10/Biden-Harris-Administrations-National-Security-Strategy-10.2022.pdf (accessed 8 December 2023).

Nye, J. 2004. *Soft Power: the Means to Success in World Politics*. New York: Public Affairs.

Nye, J. 2008. 'Public diplomacy and soft power'. *The Annals of the American Academy of Political and Social Science*, 616(1): pp. 94–109.

Nye, J. 2011. *The Future of Power*. New York: Public Affairs.

Ó Riain, S. 2004. *The Politics of High-Tech Growth: Developmental Network States in the Global Economy*. Cambridge: Cambridge University Press.

O'Neil, S. K. 2021. 'America's Supply Chains Are Foreign Policy Now'. *Foreign Policy*, 16 February. Available at: https://foreignpolicy.com/2021/02/16/biden-supply-chains-manufacturing-foreign-policy-allies-globalization/, accessed 10 December 2021.

O'Sullivan, M. L. 2013. 'The Entanglement of Energy, Grand Strategy, and International Security'. In *The Handbook of Global Energy Policy*, edited by A. Goldthau, pp. 30–48. West Sussex, UK: Wiley-Blackwell.

O'Sullivan, M. L., I. Overland, and D. Sandalow. 2017. 'The geopolitics of renewable energy'. *HKS Working Paper* No. RWP17-027, Harward.

Oberthür, S. and C. Dupont. 2021. 'The European Union's international climate leadership: Towards a grand climate strategy?'. *Journal of European Public Policy*, 28(7): pp. 1095–1114.

Oberthür, S. and M. Pallemaerts. 2010. 'The EU's Internal and External Climate Policies: An Historical Overview'. In *The New Climate Policies of the European Union: Internal Legislation and Climate Diplomacy*, edited by S. Oberthür and M. Pallemaerts, pp. 27–63. Brussels: VUB Press.

Oberthür, S. and C. Roche Kelly. 2008. 'EU leadership in international climate policy: Achievements and challenges'. *The international Spectator*, 43(3): pp. 35–50.

OECD. 2021. 'OECD 2021 Green Recovery Database'. Available at: https://www.oecd.org/coronavirus/en/themes/green-recovery (accessed 5 January 2022).

OECD-DAC. 2022. *Report on the Implementation of the DAC Recommendation on Untying Official Development Assistance*. Paris: OECD.

OPIC. 2013. 'OPIC Annual Report 2013'. Available at: https://www.dfc.gov/media/reports/archived (accessed 2 August 2023).

OPIC. 2015. 'OPIC Annual Report 2015'. Available at: https://www.dfc.gov/media/reports/archived (accessed 2 August 2023).

Opsal, R. 2019. *American and Chinese Energy Security. A Grand Strategic Approach*. London: Rowman & Littlefield.

Overland, I. 2019. 'The geopolitics of renewable energy: Debunking four emerging myths'. *Energy Research & Social Science*, 49: pp. 36–40.

Pan, J., W. Peng, M. Li, X. Wu, L. Wan, H. Zerriff, B. Elias, C. Zhang, and D. Victor. 2006. 'Rural Electrifcation in China 1950–2004. Historical Forces and Key Driving Processes'. *Program on Energy and Sustainable Development*, Working Paper No. 60, Stanford University.

Paraguassu, L. and A. Boadle. 2023. 'China and Brazil look to set up green investment fund'. *Reuters*, 24 March 2023. Available at: https://www.reuters.com/business/environment/china-brazil-look-set-up-green-investment-fund-say-lula-aides-2023-03-24/ (accessed 6 May 2023).

Parfomak, P. W., R. Campbell, R. Pirog, M. Ratner, P. Brown, J. Frittelli, and M. Humphries. 2017. 'Cross-Border Energy Trade in North America: Present and Potential'. *Congressional*

Research Service Report. Available at: https://sgp.fas.org/crs/misc/R44747.pdf (accessed 2 October 2023).

Parks, L., D. Della Porta, and M. Portos. 2023. 'Environmental and Climate Activism and Advocacy in the EU'. In *Handbook on European Union Climate Change Policy and Politics*, edited by T. Rayner, K. Szulecki, A. J. Jordan, and S. Oberthür, pp. 98–113. Cheltenham: Edward Elgar.

Prontera, A. 2017. *The New Politics of Energy Security in the European Union and Beyond. States, Markets, Institutions.* London: Routledge.

Prontera, A. 2019. *Beyond the EU Regulatory State. Energy Security and the Eurasian Gas Market*, London: ECPR Press-Rowman & Littlefield.

Prontera, A. 2020. 'Beyond the regulatory state: Rethinking energy security governance and politics in the European Union'. *Comparative European Politics*, 18(3): pp. 330–362.

Prontera, A. 2021. 'The dismantling of renewable energy policy in Italy'. *Environmental Politics*, 30(7): pp. 1196–1216.

Prontera, A. and P. Plenta. 2020. 'Catalytic Power Europe and gas infrastructural policy in the Visegrad countries'. *Energy Policy*, 137, 111189.

Prontera, A. and R. Quitzow. 2022. 'The EU as catalytic state? Rethinking European climate and energy governance'. *New Political Economy*, 27(3): pp. 517–531.

Prontera, A. and R. Quitzow. 2023. 'Catalytic Power Europe: Blended finance in European external action'. *Journal of Common Market Studies*, 61(4): pp. 988–1006

Prontera, A. 2024. 'Winter Is Coming: Russian Gas, Italy and the Post-war European Politics of Energy Security'. *West European Politics*, 47(2): pp. 382–407.

Qi, J. J. and P. Dauvergne. 2022. 'China's rising influence on climate governance: Forging a path for the global South'. *Global Environmental Change*, 73, 102484.

Qi, Y. and T. Wu. 2013. 'The politics of climate change in China'. *Wiley Interdisciplinary Reviews: Climate Change*, 4(4): pp. 301–313.

Qin, Y. 2014. 'Continuity through change: Background knowledge and China's international strategy'. *Chinese Journal of International Politics*, 7(3): pp. 285–314.

Quitzow, R., G. Bersalli, L. Eicke, J. Jahn, J. Lilliestam, F. Lira, and B. Xue. 2021. 'The COVID-19 crisis deepens the gulf between leaders and laggards in the global energy transition'. *Energy Research & Social Science*, 74, 101981.

Quitzow, R., G. Bersalli, J. Lilliestam, and A. Prontera. 2023. 'Green Recovery: Catalyst for an Enhanced EU Role in Climate and Energy Policy?'. In *Handbook on European Union Climate Change Policy and Politics*, edited by T. Rayner, K. Szulecki, A. J. Jordan, and S. Oberthür, pp. 351–366. Cheltenham: Edward Elgar.

Quitzow, R. and X. Gong. 2023. 'The Geopolitics of Hydrogen: The Emerging Role of China'. *Oxford Energy Forum*, Issue 137, August 2023, pp. 23–28. Oxford: Oxford Institute for Energy Studies. Available at: https://www.oxfordenergy.org/wpcms/wp-content/uploads/2023/08/OEF-137.pdf (accessed 26 July 2024).

Quitzow, R. and S. Thielges. 2022. 'The German energy transition as soft power'. *Review of International Political Economy*, 29(2): pp. 598–623

Ramseur, J. L. 2023. 'Inflation Reduction Act of 2022 (IRA): Provisions Related to Climate Change'. *Congressional Research Service Report.* Available at: https://crsreports.congress.gov/product/pdf/R/R47262 (accessed 4 December 2023).

Randall, S. J. 2005. *United States Foreign Oil Policy since World War I: For Profits and Security.* Montreal: McGill-Queen's Press-MQUP.

Rauchfleisch, A. and M. S. Schäfer. 2018. 'Climate change politics and the role of China: A window of opportunity to gain soft power?'. *International Communication of Chinese Culture*, 5: pp. 39–59.

Rayner, T. and A. Jordan. 2016. 'Climate change policy in the European Union'. *Oxford Research Encyclopaedias of Climate Science*. Available at: https://doi.org/10.1093/acrefore/9780190228620.013.47 (accessed 4 June 2023).

Regilme, S. 2023. 'United States foreign aid and multilateralism under the Trump presidency'. *New Global Studies*, 17(1): pp. 45–69.

REN21. 2021. 'Renewables Global Status Report 2021 Data Pack'. Available at: https://www.ren21.net/wp-content/uploads/2019/05/GSR2021_Data_Pack.xlsx (accessed 6 October 2022).

REN21. 2022. 'Renewables Global Status Report 2022'. Available at: https://www.ren21.net/wp-content/uploads/2019/05/GSR2022_Full_Report.pdf (accessed 22 January 2023).

REN21. 2023. 'Renewables Global Status Report 2023 Data Pack'. Available at: https://www.ren21.net/reports/global-status-report/ (accessed 22 August 2023).

Reuters. 2023a. 'World Bank, multilateral development banks jointly seek to boost lending power'. *Reuters*, 13 October 2023. Available at: https://www.reuters.com/markets/world-bank-multilateral-development-banks-jointly-seek-boost-lending-power-2023-10-13/ (accessed 2 November 2023).

Reuters. 2023b. 'China Energy to build green hydrogen project in Egypt with expected investments worth $5.1 bln'. *Reuters*, 8 March 2023. Available at: https://www.reuters.com/article/egypt-china-greenhyrogen-idUSS8N35D0E9 (accessed 29 August 2023).

Ritchie, H., M. Roser, and P. Rosado. 2023. 'Energy Data'. Available at: https://ourworldindata.org/energy (accessed 4–6 September 2023).

Robinson, D. 2013. 'US Energy and Climate Change Policies. Obama's Second Term'. *Oxford Energy Comment*, February 2013. Oxford: Oxford Institute for Energy Studies. Available at: https://www.oxfordenergy.org/wpcms/wp-content/uploads/2013/03/US-energy-and-climate-change-policies-Obamas-Second-Term.pdf (accessed 25 July 2024).

Rodrik, D. 2014. 'Green industrial policy'. *Oxford Review of Economic Policy*, 30(3): pp. 469–491.

Roland, A. and P. Shiman. 2002. *Strategic computing: DARPA and the quest for machine intelligence, 1983–1993*. Cambridge, MA: MIT Press.

Rolley, S. 2023. 'Congo to hike stake in copper, cobalt venture with China'. *Reuters*, 24 May 2023. Available at: https://www.reuters.com/markets/commodities/congo-hike-stake-copper-cobalt-venture-with-china-2023-05-24/ (accessed 3 August 2023).

Roy, A. 2024. 'Tesla CEO Musk: Chinese EV firms will "demolish" rivals without trade barriers'. *Reuters*, 25 January 2024. Available at: https://www.reuters.com/business/autos-transportation/tesla-ceo-musk-chinese-ev-firms-will-demolish-rivals-without-trade-barriers-2024-01-25/ (accessed 3 February 2024).

Ruggie, J. G. 1992. 'Multilateralism: The anatomy of an institution'. *International Organization*, 46(3): pp. 561–598.

Sabyrbekov, R. and I. Overland. 2024. 'Small and large friends of the EU's carbon border adjustment mechanism: Which non-EU countries are likely to support it?'. *Energy Strategy Reviews*, 51, 101303.

Sandalow, D., M. Meidan, P. Andrews-Speed, A. Hove, S. Yue Qiu, and E. Downie. 2022. *Guide to Chinese Climate Policy 2022*. Oxford: Oxford Institute for Energy Studies.

Sattich, T., D. Freeman, D. Scholten, and S. Yan. 2021. 'Renewable energy in EU-China relations: Policy interdependence and its geopolitical implications'. *Energy Policy*, 156, 112456.

Schimmelfennig, F. and U. Sedelmeier. 2004. 'Governance by conditionality: EU rule transfer to the candidate countries of Central and Eastern Europe'. *Journal of European Public Policy*, 11(4): pp. 661–679.

Scholten, D. 2018. *The Geopolitics of Renewables*. Cham: Springer.

Scholten, D. (ed). 2023. *Handbook on the Geopolitics of the Energy Transition.* Cheltenham: Edward Elgar.

Scholten, D., M. Bazilian, I. Overland, and K. Westphal. 2020. 'The geopolitics of renewables: New board, new game'. *Energy Policy*, 138, 111059.

Scholten, D. and R. Bosman. 2016. 'The geopolitics of renewables: Exploring the political implications of renewable energy systems'. *Technological Forecasting and Social Change*, 103: pp. 273–283.

Schubert, S. R., J. Pollak, and M. Kreutler. 2016. *Energy Policy of the European Union.* London: Palgrave Macmillan.

SCO. 2021. 'Joint Communique following the 20th Meeting of the Council of Heads of Government (Prime Ministers) of the Member States of the Shanghai Cooperation Organisation'. Available at: http://eng.sectsco.org/documents/ (accessed 15 June 2023).

SCO. 2022a. 'Statement by the Shanghai Cooperation Organisation Heads of State Council on Climate Change Response'. Available at: http://eng.sectsco.org/documents/ (accessed 3 July 2023).

SCO. 2022b. 'Statement of the SCO Heads of State Council on Ensuring Reliable, Sustainable and Diversified Supply Chains'. Available at: http://eng.sectsco.org/documents/ (accessed 3 July 2023).

Seaman, J. 2020. 'China and the New Geopolitics of Technical Standardization'. *Notes de l'IFRI*, Paris. Available at: https://www.ifri.org/sites/default/files/atoms/files/seaman_china_standardization_2020.pdf (accessed 8 April 2023).

Shaffer, G. and H. Gao. 2020. 'A new Chinese economic order?'. *Journal of International Economic Law*, 23(3): pp. 607–635.

Shinn, D. H. 2016. 'The environmental impact of China's investment in Africa'. *Cornell International Law Journal*, 49(1): pp. 25–68

Siddi, M. 2023. *European Energy Politics. The Green Transition and EU-Russia Energy Relations.* Cheltenham: Edward Elgar.

Siddi, M. and I. Kustova. 2021. 'From a liberal to a strategic actor: The evolution of the EU's approach to international energy governance'. *Journal of European Public Policy*, 28(7): pp. 1076–1094.

Siddi, M. and F. Prandin. 2023. 'Governing the EU's Energy Crisis: The European Commission's Geopolitical Turn and its Pitfalls'. *Politics and Governance*, 11(4): pp. 1–11.

Simpson, M. M. 2001. 'Climate Change: Federal Research, Technology, and Related Programs'. *Congressional Research Service Report.* Available at: https://www.everycrsreport.com/files/20011003_RL30452_01c231c4349dffa4fb569d5d1e85dfd5ab43290e.pdf (accessed 10 January 2024).

Skjærseth, J. B. and P. O. Eikeland. 2023. 'Governing EU Low-Carbon Innovation: From Strategic Energy Technology Plan to European Green Deal'. In *Handbook on European Union Climate Change Policy and Politics*, edited by T. Rayner, K. Szulecki, A. J. Jordan, and S. Oberthür, pp. 259–274. Cheltenham: Edward Elgar.

Skovgaard, J. 2014. 'EU climate policy after the crisis'. *Environmental Politics*, 23(1): 1–17.

Slaughter, A-M. 2011. 'A New Theory for the Foreign Policy Frontier: Collaborative Power'. *The Atlantic.* Available at: http://www.theatlantic.com/international/archive/2011/11/a-new-theory-for-the-foreignpolicy-frontier-collaborative-power/249260/ (accessed 14 February 2022).

Slaughter, A-M. 2017. *The Chessboard and the Web: Strategies of Connection in a Networked World.* New Haven, CT: Yale University Press.

Solorio, I. and P. Bocquillon. 2017. 'EU Renewable Energy Policy: A Brief Overview of Its History and Evolution'. In *A Guide to EU Renewable Energy Policy. Comparing Europeanization and Domestic Policy Change in EU Member States*, edited by I. Solorio and H. Jörgens, pp. 23–42. Cheltenham: Edward Elgar.

Song, D., B. Jia, and H. Jiao. 2022. 'Review of Renewable Energy Subsidy System in China'. *Energies*, 15(19), 7429.

Sovacool, B. K. 2008. *The Dirty Energy Dilemma: What's Blocking Clean Power in the United States*. Westport, CT: Praeger.

Stanicek, B. and S. Tarpova. 2022. *China's strategic interests in the Western Balkans*. European Parliamentary Research Service, Briefing June 2022, Strasburg.

Stensdal, I. 2014. 'Chinese climate-change policy, 1988–2013: moving on up'. *Asian Perspective*, 38(1): pp. 111–135.

Stokes, D. and S. Raphael. 2010. *Global Energy Security and American Hegemony*. Baltimore: Johns Hopkins University Press.

Stokes, L. C. and H. L. Breetz. 2018. 'Politics in the US energy transition: Case studies of solar, wind, biofuels and electric vehicles policy'. *Energy Policy*, 113: pp. 76–86.

Stopford, J. and S. Strange. 1991. *Rival States, Rival Firms: Competition for World Market Shares*. Cambridge: Cambridge University Press.

Strupczewski, J. 2023. 'Seven EU countries oppose new EU funding as response to U.S. subsidy plan'. *Reuters*, 27 January 2023. Available at: https://www.reuters.com/markets/europe/seven-eu-countries-oppose-new-eu-funding-response-us-subsidy-plan-letter-2023-01-27/ (accessed 4 March 2023).

Sutter, R. G. 2016. *Chinese Foreign Relations: Power and Policy Since the Cold War*. New York: Rowman & Littlefield.

Szulecki, K. and I. Overland. 2020. 'Energy democracy as a process, an outcome and a goal: A conceptual review'. *Energy Research & Social Science*, 69, 101768.

Tagliapietra, S., C. Trasi, and R. Veugelers. 2023. 'Europe's green industrial policy'. In *Sparking Europe's New Industrial Revolution. A Policy for Net Zero, Growth and Resilience*, edited by S. Tagliapietra and R. Veugelers, pp. 166–186. Brussels: Blueprint Series N. 33, Brugel.

Tagliapietra, S. and R. Veugelers. 2023. *Sparking Europe's New Industrial Revolution. A Policy for Net Zero, Growth and Resilience*. Brussels: Blueprint Series N. 33, Brugel.

Tagliapietra, S., R. Veugelers, and J. Zettelmeyer. 2023. 'Rebooting the European Union's Net Zero Industry Act'. *Brugel Policy Brief*. Available at: https://www.bruegel.org/policy-brief/rebooting-european-unions-net-zero-industry-act (accessed 3 January 2024).

Tago, A. 2017. 'Multilateralism, Bilateralism, and Unilateralism in Foreign Policy'. *Oxford Research Encyclopaedia of Politics*. Oxford: Oxford University Press.

Tan, X., Y. Zhao, C. Polycarp, and J. Bai. 2013. 'China's overseas investments in the wind and solar industries: trends and drivers'. *World Resources Institute Working Paper*, Washington, DC.

Tennant, C. 2023. 'Lithium's green potential fails to defuse Europe's opposition to mining'. *Euronews*, 23 February 2023. Available at: https://www.euronews.com/my-europe/2023/02/23/lithiums-green-potential-fails-to-defuse-europes-opposition-to-mining (accessed 5 January 2024).

Terzi, A., M. Sherwood, and A. Singh. 2023. 'European industrial policy for the green and digital revolution'. *Science and Public Policy*, 50(5): pp. 842–857.

Thurbon, E. and L. Weiss. 2021. 'Economic statecraft at the frontier: Korea's drive for intelligent robotics'. *Review of International Political Economy*, 28(1): pp. 103–127.

Tian, J., L. Yu, R. Xue, S. Zhuang, and Y. Shan. 2021. 'Global low-carbon energy transition in the post-COVID-19 era'. *Applied Energy*, 307, 118205.

Tocci, N. 2023. *A Green and Global Europe*. Cambridge: Polity Press.

Tritto, A. 2023. 'How Indonesia Used Chinese Industrial Investments to Turn Nickel into the New Gold'. Carnegie Endowment for International Peace, April 2023, Washington, DC.

Tsang, B., B. Schäpe, and A. Hackbarth. 2023. 'Follow the Money. Chinese Climate-Related Finance to the Global South'. *E3G Briefing Paper*, April 2023. Available at: https://www.e3g.org/wp-content/uploads/E3G-Briefing-Follow-the-Money-Chinese-climate-related-finance-to-the-Global-South.pdf (accessed 5 August 2023).

Tunsjø, Ø. 2013. *Security and Profit in China's Energy Policy: Hedging against Risk*. New York: Columbia University Press.

Tyson, A., C. Funk, and B. Kennedy. 2023. 'What the data says about Americans' views of climate change'. Pew Research Center, 9 August 2023. Available at: https://www.pewresearch.org/short-reads/2023/08/09/what-the-data-says-about-americans-views-of-climate-change/#:~:text=By%20contrast%2C%20about%20one%2Din,across%2019%20countries%20in%202022 (accessed 4 December 2023).

Underhill, G. R. 2000. 'State, market, and global political economy: genealogy of an (inter-?) discipline'. *International Affairs*, 76(4): pp. 805–824.

UNFCC. 2023. 'COP28 Agreement Signals "Beginning of the End" of the Fossil Fuel Era'. Available at: https://unfccc.int/news/cop28-agreement-signals-beginning-of-the-end-of-the-fossil-fuel-era (accessed 28 December 2023).

Urpelainen, J. and T. Van de Graaf. 2015. 'Your place or mine? Institutional capture and the creation of overlapping international institutions'. *British Journal of Political Science*, 45(4): pp. 799–827.

Urpelainen, J. and T. Van de Graaf. 2018. 'United States non-cooperation and the Paris agreement'. *Climate Policy*, 18(7): 839–851.

US Department of Commerce. 2019. 'A Federal Strategy to Ensure Secure and Reliable Supplies of Critical Minerals'. Available at: https://www.commerce.gov/sites/default/files/2020-01/Critical_Minerals_Strategy_Final.pdf (accessed 4 October 2023).

US Department of Defence. 2020. 'DOD Announces Rare Earth Element Awards to Strengthen Domestic Industrial Base'. 17 November 2020. Available at: https://www.defense.gov/News/Releases/release/article/2418542/dod-announces-rare-earth-element-awards-to-strengthen-domestic-industrial-base/ (accessed 6 December 2023).

US Department of Defence. 2022. 'United States Army Climate Strategy'. Available at: https://www.army.mil/e2/downloads/rv7/about/2022_army_climate_strategy.pdf (accessed 6 October 2023).

US Department of Energy. 2024. 'Building America's Clean Energy Future'. Available at: https://www.energy.gov/invest (accessed 8 February 2024).

US Department of State. 2008. 'United States Climate Change Actions'. Available at: https://2001-2009.state.gov/documents/organization/112549.pdf (accessed 23 December 2023).

US Department of State. 2016. 'Overview of the Global Climate Change Initiative'. *U.S. Climate Finance 2010–2015*. Available at: https://2009-2017.state.gov/documents/organization/250737.pdf (accessed 4 November 2023).

US Department of State. 2019. 'A Free and Open Indo-Pacific. Advancing a Shared Vision'. Available at: https://www.state.gov/wp-content/uploads/2019/11/Free-and-Open-Indo-Pacific-4Nov2019.pdf (accessed 6 October 2023).

US Department of State. 2023a. 'Sunnylands Statement on Enhancing Cooperation to Address the Climate Crisis'. 14 November 2023. Available at: https://www.state.gov/sunnylands-statement-on-enhancing-cooperation-to-address-the-climate-crisis/ (accessed 5 January 2024).

US Department of State. 2023b. 'Progress Report on President Biden's Climate Finance Pledge'. Available at: https://www.state.gov/progress-report-on-president-bidens-climate-finance-pledge/#:~:text=That%20is%20why%20President%20Biden,provided%20by%20the%20United%20States (accessed 2 January 2024).

US Department of State and Executive Office of the President. 2021. 'The Long-Term Strategy of the United States. Pathways to Net-Zero Greenhouse Gas Emissions by 2050'. Available at: https://www.whitehouse.gov/wp-content/uploads/2021/10/US-Long-Term-Strategy.pdf (accessed 23 June 2023).

US Government Accountability Office. 2014. 'DOE Loan Programs'. Available at: https://www.gao.gov/assets/gao-14-645t.pdf (accessed 5 October 2023).

US International Trade Commission. 2022. 'Chinese Wind Turbine Export Growth Continued in 2021'. Available at: https://www.usitc.gov/publications/332/executive_briefings/ebot_chinese_wind_turbine_export_growth_continued_in_2021.pdf (accessed 19 July 2024).

USAID. 2020. 'Factsheet Renewable Energy'. Available at: https://www.usaid.gov/sites/default/files/2022-05/USAID_ENRM_Factsheet_Renewable_Energy.pdf (accessed 12 September 2023).

USAID. 2022. 'Climate Strategy 2022–2030'. Available at: https://www.usaid.gov/sites/default/files/2022-11/USAID-Climate-Strategy-2022-2030.pdf (accessed 18 November 2023).

USAID. 2023. 'Interactive Map of USAID-Supported Renewable Energy Auctions'. Available at: https://storymaps.arcgis.com/stories/913c85b784124906825898c5f85da330 (accessed 29 December 2023).

USTDA. 2018. 'Final Overview of the 1603 Program'. Available at: https://home.treasury.gov/system/files/276/P-Status-overview-2018-03-01.pdf (accessed 3 September 2023).

Vakulchuk, R., I. Overland, and D. Scholten. 2020. 'Renewable energy and geopolitics: A review'. *Renewable and Sustainable Energy Reviews*, 122, 109547.

Velut, J. B., D. Baeza-Breinbauer, M. De Bruijne, E. Garnizova, M. Jones, K. Kolben, L. Oules, V. Rousas, F. Tigere Pittet, and T. Zamparutti. 2022. 'Comparative analysis of trade and sustainable development provisions in free trade agreements'. *LSE Trade Policy Hub*. Available at: https://www.lse.ac.uk/business/consulting/assets/documents/TSD-Final-Report-Feb-2022.pdf (accessed 25 May 2023).

Vivoda, V. 2010. 'International Oil Companies, US Government and Energy Security Policy: An Interest-Based Analysis'. *International Journal of Global Energy Issues*, 33(1–2): pp. 73–88.

Vogel, D. 1995. 'Trading Up'. *Consumer and Environmental Protection in a Global Economy*. Cambridge, MA: Harvard University Press.

Vogler, J. and C. Bretherton. 2006. 'The European Union as a protagonist to the United States on climate change'. *International Studies Perspectives*, 7(1): pp. 1–22.

Voituriez, T., W. Yao, and M. L. Larsen. 2019. 'Revising the 'host country standard' principle: a step for China to align its overseas investment with the Paris Agreement'. *Climate Policy*, 19(10): pp. 1205–1210.

Wälde, T. W. 2008. 'US Foreign Oil Policy since World War I: For Profits and Security'. *Journal of World Energy Law & Business*, 1(1): pp. 113–117.

Wallace, W. L., L. Jingming, and G. Shangbin. 1998. 'The Use of Photovoltaics for Rural Electrification in North-western China'. In *Preprints for the 2nd World Conference on Photovoltaic Solar Energy Conversion*, pp. 131–136, Vienna, Austria.

Wang, B. 2018. 'After China's ministerial shake-up, what's next for South-South climate cooperation?'. *China Dialogue*, 19 June. Available at: https://chinadialogue.net/en/climate/10685-after-china-s-ministerial-shake-up-what-s-next-for-south-south-climate-cooperation/ (accessed 5 may 2023).

WBIF. 2021. 'Western Balkans Investment Framework. Annual Report 2020'. Available at: https://wbif.eu/storage/app/media/Library/4.%20Annual%20Report/WBIF%20Annual%20Report%202020/WBIF%20Annual%20Report%202020.pdf (accessed 8 May 2023).

Weber, B. 2017. 'The European Neighbourhood Policy and energy'. In *The Routledge Handbook on the European Neighbourhood Policy*, edited by T. Schumacher, A. Marchetti and T. Demmelhuber, pp. 381–392. London: Routledge.

Weiss, L. 1998. *The Myth of the Powerless State: Governing the Economy in a New Era.* Ithaca, NY: Cornell University Press.

Weiss, L. 2003. 'Guiding Globalisation in East Asia: New Roles for Old Developmental States'. In *States in the Global Economy: Bringing Domestic Institutions Back In*, edited by L. Weiss, pp. 245–270. Cambridge: Cambridge University Press.

Weiss, L. 2014. *America Inc.? Innovation and Enterprise in the National Security State.* Ithaca, NY: Cornell University Press.

Weiss, L. 2021. 'Re-emergence of great power conflict and US economic statecraft'. *World Trade Review*, 20(2): pp. 152–168.

Weiss, L. and E. Thurbon. 2021. 'Developmental state or economic statecraft? Where, why and how the difference matters'. *New Political Economy*, 26 (3): pp. 472–489.

Westphal, K., M. Pastukhova, and J. M. Pepe. 2022. 'Geopolitics of Electricity: Grids, Space and (political) Power'. SWP Research Paper 6, March 2022, Berlin.

Wettestad, J. 2023. 'Proactive prevention of carbon leakage? The EU Carbon Border Adjustment Mechanism'. In *Handbook on European Union Climate Change Policy and Politics*, edited by T. Rayner, K. Szulecki, A. J. Jordan, and S. Oberthür, pp. 231–245. Cheltenham: Edward Elgar.

The White House. 2011a. 'Remarks by the President in State of Union Address'. 25 January 2011. Available at: https://obamawhitehouse.archives.gov/the-press-office/2011/01/25/remarks-president-state-union-address (accessed 3 December 2023).

The White House. 2011b. 'A Strategy for American Innovation: Securing Our Economic Growth and Prosperity'. Available at: https://obamawhitehouse.archive.gov/sites/default/files/uploads/InnovationStrategy.pdf (accessed 13 September 2023).

The White House. 2014. 'The All-of-the-Above Energy Strategy as a Path to Sustainable Economic Growth'. 29 May 2014. Available at: https://obamawhitehouse.archives.gov/blog/2014/05/29/new-Report-all-above-energy-strategy-path-sustainable-economic-growth (accessed 5 October 2023).

The White House. 2015. 'A Strategy for American Innovation'. Available at: https://obamawhitehouse.archives.gov/sites/default/files/strategy_for_american_innovation_october_2015.pdf (accessed 5 October 2023).

The White House. 2016. 'The Recovery Act Made the Largest Single Investment In Clean Energy In History'. 25 February 2016. Available at: https://obamawhitehouse.archives.gov/the-press-office/2016/02/25/fact-sheet-recovery-act-made-largest-single-investment-clean-energy (accessed 23 October 2023).

The White House. 2018. 'Strategy for American Leadership in Advanced Manufacturing'. Available at: https://trumpwhitehouse.archives.gov/wp-content/uploads/2018/10/Advanced-Manufacturing-Strategic-Plan-2018.pdf (accessed 5 December 2023).

The White House. 2021. 'Building Resilient Supply Chains, Revitalizing American Manufacturing, and Fostering Broad-Based Growth'. Available at: https://www.whitehouse.gov/wp-content/uploads/2021/06/100-day-supply-chain-review-report.pdf (accessed 3 December 2023).

The White House. 2022a. 'CHIPS and Science Act Will Lower Costs, Create Jobs, Strengthen Supply Chains, and Counter China'. 9 August 2022. Available: https://www.whitehouse.gov/briefing-room/statements-releases/2022/08/09/fact-sheet-chips-and-science-act-will-lower-costs-create-jobs-strengthen-supply-chains-and-counter-china/ (accessed 23 December 2023).

The White House. 2022b. 'President Biden Takes Bold Executive Action to Spur Domestic Clean Energy Manufacturing'. 6 June 2022. Available at: https://www.whitehouse.gov/briefing-room/statements-releases/2022/06/06/fact-sheet-president-biden-takes-bold-executive-action-to-spur-domestic-clean-energy-manufacturing/ (accessed 22 December 2023).

The White House. 2022c. 'US Strategy Toward Sub-Saharan Africa'. Available at: https://www.whitehouse.gov/wp-content/uploads/2022/08/U.S.-Strategy-Toward-Sub-Saharan-Africa-FINAL.pdf (accessed 29 December 2023).

The White House. 2022d. 'Indo-Pacific Strategy of the United States'. Available at: https://www.whitehouse.gov/wp-content/uploads/2022/02/U.S.-Indo-Pacific-Strategy.pdf (accessed 11 October 2023).

The White House. 2022e. 'US-UAE Partnership to Accelerate Transition to Clean Energy'. 1 November 2022. Available at: https://www.whitehouse.gov/briefing-room/statements-releases/2022/11/01/fact-sheet-u-s-uae-partnership-to-accelerate-transition-to-clean-energy-pace/ (19 December 2023).

The White House. 2022f. 'Vice President Harris Launches New Initiatives to Strengthen US-Philippines Alliance'. 20 November 2022. Available at: https://www.whitehouse.gov/briefing-room/statements-releases/2022/11/20/fact-sheet-vice-president-harris-launches-new-initiatives-to-strengthen-u-s-philippines-alliance/ (accessed 10 January 2024).

The White House. 2023a. 'Remarks by National Security Advisor Jake Sullivan on Renewing American Economic Leadership at the Brookings Institution'. 27 April 2023. Available at: https://www.whitehouse.gov/briefing-room/speeches-remarks/2023/04/27/remarks-by-national-security-advisor-jake-sullivan-on-renewing-american-economic-leadership-at-the-brookings-institution/ (accessed 6 October 2023).

The White House. 2023b. 'United States Government National Standards Strategy for Critical and Emerging Technology'. Available at: https://www.whitehouse.gov/wp-content/uploads/2023/05/US-Gov-National-Standards-Strategy-2023.pdf (accessed 8 October 2023).

The White House. 2023c. 'Vice President Harris Announces Initiatives to Deepen the U.S. Partnership with Tanzania'. 30 March 2023. Available at: https://www.whitehouse.gov/briefing-room/statements-releases/2023/03/30/fact-sheet-vice-president-harris-announces-initiatives-to-deepen-the-u-s-partnership-with-tanzania-2 (accessed 4 December 2023).

The White House. 2023d. 'Australia-United States Climate, Critical Minerals and Clean Energy Transformation Compact'. 20 May 2023. Available at: https://www.whitehouse.gov/briefing-room/statements-releases/2023/05/20/australia-united-states-climate-critical-minerals-and-clean-energy-transformation-compact/#:~:text=Australia%20and%20the%20United%20States%20affirm%20a%20shared%20view%20of,%2C%20diplomacy%2C%20and%20development%20counterparts (accessed 8 December 2023).

The White House. 2024. 'Budget of the US Government. Fiscal Year 2024'. Available at: https://www.whitehouse.gov/wp-content/uploads/2023/03/budget_fy2024.pdf (accessed 5 February 2024).

USGS. 2022. 'Mineral Commodity Summaries 2022 – Rare-Earths'. *US Geological Survey*. Available at: https://pubs.usgs.gov/periodicals/mcs2022/mcs2022-rare-earths.pdf (accessed 18 July 2024).

Wigell, M., S. Scholvin, and M. Aaltola. 2019. *Geo-Economics and Power Politics in the 21st Century: The Revival of Economic Statecraft.* London: Routledge.

Woo-Cumings, M. (ed.). 1999. *The Developmental State.* Ithaca, NY: Cornell University Press.

Wouters, L. 2023. 'Key Players: Why Mining Is Central to the EU's Critical Raw Materials Ambitions in Africa'. ECFR Commentaries, 24 March 2023. Available at: https://ecfr.eu/article/key-players-why-mining-is-central-to-the-eus-critical-raw-materials-ambitions-in-africa/ (accessed 5 November 2023).

Wright, T. 2017. *All Measures Short of War: The Contest for the Twenty-First Century and the Future of American Power.* New Haven: Yale University Press.

Wu, F. 2016. 'Shaping China's climate diplomacy: wealth, status, and asymmetric interdependence'. *Journal of Chinese Political Science*, 21, pp. 199–215.

Wu, S. and A. Chong. 2018. 'Developmental rail politics: The political economy of China's high-speed rail projects in Thailand and Indonesia'. *Contemporary Southeast Asia: A Journal of International and Strategic Affairs*, 40(3): pp. 503–526.

Wübbeke, J., M. Meissner, M. J. Zenglein, J. Ives, and B. Conrad. 2016. 'Made in China 2025. The making of a high-tech superpower and consequences for industrial countries'. *Merics Papers on China*, No. 2/2016, Mercator Institute for China Studies.

Wurzel, R. K. and J. Connelly (eds.). 2010. *The European Union as a Leader in International Climate Change Politics.* London: Routledge.

Wurzel, R. K., J. Connelly, and D. Liefferink (eds.). 2016. *The European Union in International Climate Change Politics. Still Taking a Lead?.* London: Routledge.

Xu, C. 2011. 'The Fundamental Institutions of China's Reforms and Development'. *Journal of Economic Literature*, 49(4): pp. 1076–1151.

Yang, H., B. A. Simmons, R. Ray, C. Nolte, S. Gopal, Y. Ma, X. Ma, and K. P. Gallagher. 2021. 'Risks to global biodiversity and Indigenous lands from China's overseas development finance'. *Nature Ecology & Evolution*, 5(11): pp. 1520–1529.

Yang, J. 2022. 'Understanding China's changing engagement in global climate governance: a struggle for identity'. *Asia Europe Journal*, 20: pp. 357–376.

Yeophantong, P. and E. Goh. 2022. 'China as a 'Partial' Environmental Great Power'. In *Great Powers, Climate Change and Global Environmental Responsibilities*, edited by R. Falkner and B. Buzan, pp. 71–95. Oxford: Oxford University Press.

Yergin, D. 2006. 'Ensuring Energy Security'. *Foreign Affairs*, 85(2): pp. 69–82.

Young, O. R. 1991. 'Political Leadership and Regime Formation: On the Development of Institutions in International Society'. *International Organization*, 45(3): pp. 281–309

Youngs, R. 2013. 'The EU's Global Climate and Energy Policies: Gathering or Losing Momentum?'. In *The Handbook of Global Energy Policy*, edited by A. Goldthau, pp. 421–434. Sussex, UK: Wiley-Blackwell.

Zabanova, Y. 2023. 'The EU in the Global Hydrogen Race: Bringing together Climate Action, Energy Security, and Industrial Policy'. *RIFS Discussion Paper*, December 2023, https://doi.org/10.48481/rifs.2023.034

Zhang, F. 2018. 'The Chinese developmental state: standard accounts and new characteristics'. *Journal of International Relations and Development*, 21: pp. 739–768.

Zhao, F., F. Bai, X. Liu, and Z. Liu. 2022. 'A Review on Renewable Energy Transition under China's Carbon Neutrality Target'. *Sustainability*, 14(22): p. 15006.

Zhao, S. (ed.). 2014. *China's Search for Energy Security: Domestic Sources and International Implications.* New York: Routledge.

Index

For the benefit of digital users, indexed terms that span two pages (e.g., 52–53) may, on occasion, appear on only one of those pages.